Decoding AI

Business operations today show evidence of profound DNA-level transformations brought about by existing artificial intelligence (AI) technology. *Decoding AI: Practical Implementations for Business Operations* reveals real artificial intelligence potentials through practical insights and innovative frameworks, and transformative case studies that operate across different business sectors.

The book explores AI model types, including large language models (LLMs) and multimodal embeddings alongside reinforcement learning and decision-priming simulations while discussing their impact on finance and marketing alongside healthcare and customer service, and fraud detection and personalized commerce functions. Through its blend of academic AI knowledge and practical business insights, the content enables readers to understand complex algorithms better while achieving improved operational facility.

Key highlights include:

- Ethical frameworks for responsible AI and human-aligned intelligence
- Data-centric strategies to improve transparency, explainability, and governance
- Domain-specific innovations in pricing, negotiation modeling, and behavioral prediction
- Real-world case studies demonstrating the transformative power of AI in decision-making
- A special focus on the integration of AI in strategic, financial, and customer-facing functions

Whether you're a C-suite executive, startup founder, researcher, academician, or tech enthusiast, this book is designed to equip you with both the strategic vision and technical foundation to embrace AI confidently in your business journey.

Let this volume be your guide in navigating the evolving AI landscape—not just to keep up with innovation, but to lead it.

Decoding AI

Practical Implementations for Business Operations

Edited by
Alexander Didenko, Hemachandran K,
Raul Villamarin Rodriguez, Andrey Korshunov,
and Maria Bernstein

Routledge
Taylor & Francis Group
A PRODUCTIVITY PRESS BOOK

Designed cover image: Web Large Image (Public)

First published 2026
by Routledge
605 Third Avenue, New York, NY 10158

and by Routledge
4 Park Square, Milton Park, Abingdon, Oxon, OX14 4RN

Routledge is an imprint of the Taylor & Francis Group, an informa business

ISBN: 978-1-032-99969-2 (hbk)
ISBN: 978-1-032-99968-5 (pbk)
ISBN: 978-1-003-60695-6 (ebk)

DOI: 10.4324/9781003606956

Typeset in Garamond
by Deanta Global Publishing Services, Chennai, India

Contents

Introduction

In today's rapidly evolving business ecosystem, artificial intelligence (AI) is not just a technological advancement—it is a strategic imperative. As organizations strive for greater efficiency, deeper insights, and enhanced decision-making, AI has emerged as a key driver of transformation across industries.

This book, *Decoding AI: Practical Implementations for Business Operations*, brings together an interdisciplinary collection of chapters that bridge the gap between AI theory and its tangible impact on business practice. From redefining organizational power structures to applying AI in dynamic pricing, financial fraud detection, and personalized customer experiences, each chapter offers a unique lens on how AI is reshaping business functions.

The book also explores the ethical, responsible, and equitable dimensions of AI adoption. Topics such as responsible AI, equitable AI design, and the use of LLMs in human-centered strategic facilitation provide critical perspectives for decision-makers who aim to adopt AI not only for performance but also for purpose.

Crucially, this volume is grounded in real-world applications. Through case studies, frameworks, and experimental research, the contributors highlight how AI technologies are being practically implemented to generate business value. Domains such as finance, healthcare, e-commerce, hospitality, and art markets are explored in depth—revealing both the opportunities and the challenges of integrating AI across diverse operational contexts.

We believe this book will serve as a valuable resource for business leaders, researchers, students, and policymakers who are seeking to navigate the complexities of AI integration with clarity, responsibility, and impact.

We extend our sincere gratitude to the contributors for their expertise and dedication, and we hope readers will find this work both practical and thought-provoking.

Between 80% and 85% of AI projects end in failure—the highest failure rate in the digital sector. This stark statistic stands in vivid contrast to the media hype of the past three years. The causes appear diverse: poor data quality, the "disease" of flashy novelties and signaling, misaligned stakeholder incentives, premature handoffs when projects fail to deliver rapid results, and resistance from both personnel and technologies during scaling. At first glance, these factors seem heterogeneous, yet we argue that there is a common root.

Theories are often dismissed outright. Once Yogi Berra famously quipped, "In theory, there is no difference between theory and practice. In practice, there is." We agree with the wording but disagree with its underlying meaning. A powerful theoretical metaphor provides an epistemological "edge"—revealing patterns and insights that a purely practical lens cannot. In fact, "nothing is more practical than a good theory"—a sentiment famously expressed by Kurt Lewin.

Enter the cognitive theory of the firm. Consider the classic observation of the ship (Hutchins 1995), which illustrates how collective cognition emerges through the coordination of individuals and tools. A ship's ability to "know" its position in space is not the product of a single individual or device but of an intricate network of human operators, navigational instruments, and procedures. The ship constructs a representational state—an understanding of the place of the ship's model inside the world's model—through distributed collaboration, where no single entity alone at any time holds the full cognitive burden. This mirrors the way firms operate: they do not merely process inputs to generate outputs but continuously construct and maintain shared cognitive states that guide decision-making. Firms, just like universities, are knowledge-producing institutions, engaged in distributed cognition and collective sense-making. In this light, every firm—be it a modern tech startup or a medieval corporation (indeed, universities themselves originated as corporations in Middle Ages)—shares a common heritage: the creation and sharing of representational states, the aggregation of decision-making processes, memory and forgetting, information processing—all of which are distributed across both human and non-human actors. This perspective finds support in ideas ranging from the attentional view of the firm (Ocasio 1997) through Masahiko Aoki's models of associative cognition (2001), to Nooteboom's insights on inter-firm learning and trust (2000), and to the exploration of tacit knowledge and distributed decision-making (Tanaka & Mol 2021).

What, then, is an AI implementation project in this framework? AI implementation is, at its core, the automation of cognitive labor, shifting how knowledge is generated, shared, and operationalized within a firm. Thus, the successful integration of AI demands a deep understanding of how humans and machines co-create cognitive value within an organization.

This breakdown in collaboration between humans and AI is precisely the phenomenon explored by Fügener et al. (2021) in their study of delegation dynamics. Their laboratory experiments, conducted in an image classification context, reveal a striking asymmetry in delegation: while an AI that delegates tasks to humans (a process termed "inversion") improves overall performance, humans delegating tasks to AI perform significantly worse. The findings suggest that subjects in the lab, due to a lack of metacognition—an accurate assessment of their own cognitive capabilities—often misjudge the difficulty of tasks and make suboptimal delegation decisions. We argue that the external validity of Fügener et al.'s work may be even greater than the authors themselves suggest. Their controlled experiments illuminate fundamental cognitive challenges that occur in many real-world scenarios.

In this sense, the findings of Fügener et al. offer critical insights into the subtle yet decisive breakdowns in human–AI collaboration that underlie many AI implementation failures. Their findings illuminate how asymmetrical delegation dynamics—rooted in divergent metacognitive capacities between humans and algorithms—create systemic misalignments. When organizations lack mechanisms to negotiate these asymmetries, cognitive breakdowns propagate across sociotechnical systems, derailing AI integration.

This insight compels us to examine metacognition at the organizational level. Just as individuals possess metacognitive abilities enabling reflection on their knowledge gaps and decision-making strategies under uncertainty, organizations develop distributed metacognitive processes. These emerge through the interplay of material artifacts, institutional ideas, and human agents. Organizations continuously engage in self-referential assessments: diagnosing deficiencies in their cognitive architectures, determining when and how to acquire new knowledge, and restructuring workflows to address emerging challenges. Crucially, initiatives to implement AI systems constitute metacognitive acts—conscious interventions to expand organizational cognition through hybridization with nonhuman intelligence.

Expanding on this perspective, frameworks such as Cynefin, CRISP-DM, and Scrum can themselves be viewed as metacognitive tools. These frameworks enable organizations to explicitly frame the nature of their environment or situation—be it simple, complicated, complex, or chaotic—and consequently determine the appropriate methods of cognitive production, whether generating knowledge, making decisions, or managing memory and forgetting. By categorizing their environment, organizations activate metacognitive processes, intentionally adapting their cognitive strategies to align with perceived environmental conditions.

Similarly, categorizing types of uncertainty—whether epistemic versus aleatoric, or structural versus parametric—also serves as a metacognitive exercise. These classifications help organizations assess the constraints that uncertainty imposes on their cognitive outputs, particularly forecasts, and their alignment with future environmental states. Recognizing and articulating the nature of uncertainty thus becomes a metacognitive act, shaping how the organization approaches knowledge generation, decision-making, and adaptation.

Viewed through this lens, many common pitfalls in AI implementation—such as inadequate, biased, or imbalanced training data; inflated initial expectations; or unclear definitions of the problem and its stakeholders—can be reinterpreted as organizational metacognitive failures. These missteps reflect an inability by the organization to accurately assess and respond to its own cognitive limitations and to adequately frame the environmental context in which AI is deployed.

Importantly, metacognition within algorithms differs fundamentally from that within organizations. Algorithmic metacognition operates through structured rule-based instructions, enabling an AI system to recognize and handle its own operational boundaries. In contrast, organizational metacognition emerges through ongoing negotiations among human and non-human agents. Within these negotiations, agents continuously evaluate their interdependencies, indispensability, and potential compromises, collectively defining the boundaries and conditions under which effective collaboration can occur.

Expanding on this perspective, frameworks such as Cynefin, CRISP-DM, and Scrum can themselves be viewed as metacognitive tools. These frameworks enable organizations to explicitly frame the nature of their environment or situation—be it simple, complicated, complex, or chaotic—and consequently determine the appropriate methods of cognitive production, whether generating knowledge, making decisions, or managing memory and forgetting. By categorizing their environment, organizations activate metacognitive processes, intentionally adapting their cognitive strategies to align with perceived environmental conditions.

Similarly, categorizing types of uncertainty—whether epistemic versus aleatoric, or structural versus parametric—also serves as a metacognitive exercise. These classifications help organizations assess the constraints that uncertainty imposes on their cognitive outputs, particularly forecasts, and their alignment with future environmental states. Recognizing and articulating the nature of uncertainty thus becomes a metacognitive act, shaping how the organization approaches knowledge generation, decision-making, and adaptation.

Viewed through this lens, many common pitfalls in AI implementation—such as inadequate, biased, or imbalanced training data; inflated initial expectations; or unclear definitions of the problem and its stakeholders—can be reinterpreted as organizational metacognitive failures. These missteps reflect an inability by the organization to accurately assess and respond to its own cognitive limitations and to adequately frame the environmental context in which AI is deployed.

Importantly, metacognition within algorithms differs fundamentally from that within organizations. Algorithmic metacognition operates through structured rule-based instructions, enabling an AI system to recognize and handle its own operational boundaries. In contrast, organizational metacognition emerges through ongoing negotiations among human and non-human agents. Within these negotiations, agents continuously evaluate their interdependencies, indispensability, and potential compromises, collectively defining the boundaries and conditions under which effective collaboration can occur.

Analyzing this negotiation through an actor–network theoretical lens reveals how AI, as a novel actor, positions itself as an obligatory point of passage, thereby redefining roles, goals, and power relations within organizational networks. The integration of AI into organizational processes does not merely automate existing tasks but also reconfigures established hierarchies and decision-making structures, intensifying the complexities inherent in human metacognition. It prompts renegotiations among stakeholders about authority, expertise, and the distribution of cognitive responsibilities.

This theoretical perspective sets the stage for Cellerin's opening chapter, "The Silent Revolution: How AI Is Rewriting the Rules of Organizational Power," which examines how AI reshapes managerial roles and redistributes organizational power from traditional hierarchical positions toward algorithmic governance. Management shifts from direct decision-makers to meta-managers overseeing AI-driven strategic and ethical compliance. Power transitions from positional authority to informational authority, emphasizing data interpretation over hierarchical status. Cellerin emphasizes the necessity of adaptive leadership and hybrid decision-making processes, highlighting AI as an active agent that redefines organizational structures, demanding new governance models, transparency, and accountability.

Further sharpening Cellerin's analysis, Davenport and Kirby's earlier framework from "Beyond Automation" (2015) outlines how humans strategically contest AI's growing dominance by reclaiming agency and redefining themselves as obligatory points of passage within AI-driven organizational contexts:

1. **Step up**: Human actors adopt strategic oversight roles, emphasizing high-level abstraction and synthesis. Managers become meta-managers who oversee AI-driven insights, reshaping their goals toward governance rather than direct decision-making. Metacognitive asymmetry here does not give algorithms a decisive edge, as identifying, evaluating, and managing uncertainty is inherently reflective and negotiated rather than purely algorithmic. Algorithms are constrained by pre-defined rules and thus struggle with structural uncertainties (e.g., regulatory or societal changes like a Bitcoin ban), even as they excel at managing parametric uncertainties (e.g., predicting specific numerical outcomes like Bitcoin prices).
2. **Step aside**: Actors leverage interpersonal and emotional intelligence, capitalizing on uniquely human skills such as empathy, creativity, and judgment. Metacognitive asymmetry provides no decisive edge for algorithms here because interpersonal and emotional intelligence inherently involves navigating ambiguous, multimodal, context-dependent human interactions—domains algorithms cannot fully interpret or codify. Didenko et al. (2025) (see Chapter 2), in their work on multimodal embeddings for negotiators, illustrate precisely this human advantage, emphasizing how human negotiators manage multimodal complexities and social subtleties, maintaining agency in contexts inaccessible to algorithmic interpretation alone.
3. **Step in**: Humans become algorithm supervisors, maintaining power by actively monitoring and correcting automated decision-making. Here, humans increasingly integrate with algorithms, forming hybrid cognitive systems that blur traditional boundaries. This "human-algorithm fusion" creates conditions for deeper interpretability and transparency, hinting towards emerging approaches in explainable AI (XAI). As human and algorithmic cognition intertwine, critical new points of power and

agency emerge, especially in relation to control over data and transparency of decision-making processes. Munerman (2025) (see Chapter 7) further highlights how textual financial analysis through XAI supports this hybridization, enabling human agents to comprehend and supervise AI's reasoning processes, enhancing organizational metacognition.

4. **Step narrowly**: Specialized expertise becomes essential as power and influence accrue to those who possess niche, non-automatable knowledge. Actors position themselves as indispensable experts, thereby securing their relevance and bargaining power within the organizational network. However, as Shabason et al. (2025) (see Chapter 12) illustrate in their analysis of the art market, even traditionally human-dominated domains are increasingly penetrated by AI. By leveraging large-scale datasets and data-centric AI (DCAI) methodologies, Shabason demonstrates how AI can reconstruct gaps in historical data, effectively "unforgetting" lost knowledge and improving predictive accuracy. This shift challenges the presumed inviolability of Step Narrowly roles, as AI expands into domains once considered beyond automation.
5. **Step forward**: Human actors assume roles in designing, refining, and advancing AI systems themselves. Power shifts toward those who control the development and implementation of AI tools, embedding human values and strategic objectives directly into algorithmic governance. This aligns with insights from Sivarajkumar et al. (2024), who examine the role of AI as a facilitator and moderator in strategic sessions. Their work highlights how AI can enhance but also challenge traditional human moderation, reinforcing the necessity of human oversight in shaping AI-driven decision-making processes. However, this also represents another attack on Step Narrowly roles, as LLMs encroach on domains requiring structured discourse facilitation. This trend is likely to accelerate with the rise of multimodal AI models, which will further blur the lines between human and algorithmic facilitation, potentially redefining the balance of cognitive agency within collaborative settings.

Veretennikov (2020) critically examine how AI redistributes economic agency on a broader labor market scale, particularly focusing on data annotators. Despite their crucial role, annotators are systematically marginalized and treated merely as disposable labor. The authors analytically demonstrate, using Shapley Values from game theory, how fair economic compensation can incentivize annotators, ultimately reinforcing AI's predictive capabilities rather than undermining them.

Shabason et al. (2025) (see Chapter 12) further illustrate this phenomenon by examining the art market—a domain traditionally resistant to algorithmic penetration. Initially considered inaccessible (Step Narrowly), AI now increasingly encroaches upon art evaluation by leveraging extensive datasets. Through data-centric AI (DCAI) methods, Shabason shows that data itself acquires agency, dynamically addressing its gaps and imperfections, enhancing predictive performance. This paradigm shift exemplifies the movement of previously niche human roles (Step Narrowly) into closer algorithm–human collaboration (Step In), empowered by XAI techniques.

Munerman (2025) (see Chapter 7) expands the role of XAI in financial contexts, emphasizing the interpretability of AI-driven financial analyses. Munerman argues that without transparency and explainability, AI systems risk generating misleading conclusions in critical financial decisions.

Finally, Didenko et al. (2025) (see Chapter 2) address the integration of human–algorithm cognitive systems in uncertain, dynamic environments, such as financial markets. This integration is characterized by algorithms explicitly revealing their reasoning processes token-by-token, which allows humans unprecedented visibility into algorithmic "thinking." This explicit "verbatim reasoning" capability fundamentally disrupts traditional organizational metacognition, enabling human agents to directly monitor and adjust algorithmic decision-making processes in real time.

Subhani and Sumalakshmi (2025) (see Chapter 14) illustrate AI's transformative role in healthcare decision-making, highlighting how predictive analytics and real-time decision support redistribute expertise and agency between medical professionals and AI systems. Singh (2025) (see Chapter 15) explores dynamic pricing in hospitality, demonstrating AI's growing authority in shaping market interactions and consumer behavior, thus redistributing economic agency from traditional human decision-makers to automated systems. Pokuri et al. (2025) (see Chapter 9) and Haseena et al. (2025) (see Chapter 8) emphasize AI's expanding role in financial fraud detection, foregrounding the critical necessity for transparency and interpretability in algorithmic governance. Oyenuga (2025) (see Chapter 13) analyzes AI-driven customer service, arguing that AI reshapes external organizational interactions, effectively redefining customer relations and influencing customer experience through automated personalization. Collectively, these contributions underscore the broad and deep reconfiguration of organizational power dynamics driven by AI integration.

References

Aoki, M. (2001). *Toward a comparative institutional analysis.* MIT Press.

Davenport, T. H., & Kirby, J. (2015). Beyond automation. *Harvard Business Review*, 93(6), 58–65.

Fügener, A., Grahl, J., Gupta, A., & Ketter, W. (2021). Cognitive challenges in human–AI collaboration: Investigating delegation choices in human-AI teams. *ACM Transactions on Economics and Computation*, 9(4), Article 23. https://doi.org/10.1145/3481628

Hutchins, E. (1995). *Cognition in the wild.* MIT Press.

Nooteboom, B. (2000). *Learning and innovation in organizations and economies.* Oxford University Press.

Ocasio, W. (1997). Towards an attention-based view of the firm. *Strategic Management Journal*, 18(S1), 187–206. https://doi.org/10.1002/(SICI)1097-0266(199707)18:1+

Sivarajkumar, S., Kelley, M., Samolyk-Mazzanti, A., Visweswaran, S., & Wang, Y. (2024). An empirical evaluation of prompting strategies for large language models in zero-shot clinical natural language processing: algorithm development and validation study. *JMIR Medical Informatics*, 12, e55318.

Tanaka, S., & Mol, A. (2021). On the materiality of AI in organizations: Tacit knowledge and the logic of delegation. *Big Data & Society*, 8(2). https://doi.org/10.1177/20539517211050668

Veretennikov, N. P. (2020). *Arctic: Transport and information and telecommunication infrastructure, economy and national security. Formation of Economic Order*, (21). Doctor of Economic Sciences, Professor, Chief Researcher, Institute of Economic Problems named after SE Luzina, Federal Research Center KSC RAS, Apatity.

Editors

Alexander Didenko is a seasoned professional with a rich background in financial markets, banking, and academia. He holds a PhD from the Plekhanov Russian University of Economics, where his research integrated behavioral economics and data science. Alexander has extensive experience in trading algorithm development and has held various executive roles in education. Currently, he leads the AI Lab at the Moscow School of Management SKOLKOVO and the Neurosciences in Management Lab at IBS-Moscow, RANEPA. His work focuses on harnessing AI and data science to drive innovation in management and decision-making, blending his expertise in economics, finance, and technology.

Hemachandran K is the Director of the AI Research Centre and Vice Dean of the School of Business at Woxsen University, India. An accomplished academic and researcher, his expertise spans Natural Language Processing, Computer Vision, and Autonomous Robotics. He is the principal investigator for multiple patented AI systems, including nano-based necrobotics and the Unified Healthcare Intelligence System (UHIS). Dr Hemachandran serves as an expert for UNESCO and the European Commission, an ATL Mentor of Change, and an advisory board member for international tech firms. He is also a prolific author, having written and edited numerous books on AI and analytics.

Raul Villamarin Rodriguez is a cognitive technologist and behavioral engineer serving as Vice President of Woxsen University, India. With a diverse educational background in cognitive science, big data, and business, his work focuses on augmenting human intelligence through technology. He has developed several patented cognitive models and systems, such as Behavioral GPS™ and the Unified Healthcare Intelligence System, documented in over 70 publications. Raul is a sought-after speaker at global forums like Web Summit and holds numerous advisory and board roles across academia, industry, and organizations like the European Commission and UNESCO, driving innovation in AI and human–technology integration.

Andrey Korshunov is a scholar and administrator in the fields of digital transformation and strategic management. He has worked at the Russian Presidential Academy of National Economy and Public Administration (RANEPA) for over twenty years and currently serves as Deputy Director of the Institute of Management and Director of Online Master's Programs. He holds a Candidate of Sciences degree in Economics and is the co-founder of a professional community dedicated to advancing online education in higher education. In addition to his academic leadership, his expertise spans product management, corporate strategy, and finance, bridging research, teaching, and practical consulting to foster innovation in both education and industries.

Maria Bernstein is a dedicated academic and administrator with over 18 years of experience at the Russian Presidential Academy of National Economy and Public Administration (RANEPA). She holds a PhD in Economics and serves as Vice Dean, Program Director, and Associate Professor. Maria is also a certified Business Simulations Trainer and a Visiting Professor at the Moscow School of Management SKOLKOVO. Her professional background includes significant industry experience in marketing and brand management roles at companies like Iceberry and NeffeTessTechnology. She specializes in strategic planning, business management, and integrating practical business simulations into executive and master's level education.

Contributors

Egor Alaev
School of Advanced Studies
Tyumen State University
Tyumen, Russia

Kande Archana
Department of CSE
Malla Reddy College of Engineering
Hyderabad, Telangana, India

Ayasha
Christ College of Science and Management
Bangalore, Karnataka, India

Thierry Cellerin
Faculty of Business Administration
Prague University of Economics and Business
Prague, Czech Republic

Prasanth Damma
School of Technology
Woxsen University
Hyderabad, Telangana, India

Alexander Didenko
Laboratory of Neurosciences in Management
RANEPA
Moscow, Russia

Maria Egorova
SelfOrb
Manama, Bahrain

Ahmed Elghandour
School of Advanced Studies
Tyumen State University
Tyumen, Russia

Evgeny Goryunov
Gaidar Institute for Economic Policy
Moscow, Russia

Mikhail Grigoryan
Department of Business & Management
HSE University
Moscow, Russia

Shaik Valli Haseena
Presidency College
Bangalore, Karnataka, India

Karri Hema Kumar
School of Business
Woxsen University
Hyderabad, Telangana, India

Kirill Ivanov
SelfOrb
Manama, Bahrain

Yana Ivanova
Sber
Moscow, Russia

Neha Jaswani
Presidency College
Bangalore, Karnataka, India

Solomon Jeresa
University of Huddersfield
Huddersfield, United Kingdom

Artem Kraevskiy
Higher School of Economics
National Research University
Russia

Mikayil Majidov
Research Scholar
Russian Presidential Academy of National Economy and Public Administration
Moscow, Russia

Sergei Maksimov
SelfOrb
Manama, Bahrain

Julia Malovitsa
SelfOrb
Manama, Bahrain

Mark Mamrykin
Skolkovo School of Management
Moscow, Russia

Yana Milenkaya
Laboratory of Neurosciences in Management
Russian Presidential Academy of National Economy and Public Administration
Moscow, Russia

Jamalnake Shaik Mohammed Hafeez
School of Technology
Woxsen University
Hyderabad, Telangana, India

Ilia Munerman
R & D Head
Interfax Information Group
Moscow, Russia

Ram Kumar Nimmakayala
AI, Western Governors University
Millcreek, Utah, United States

Sunday Alewo Omale
Federal University Gashua
Yobe State, Nigeria

Oyenuga Michael Oyedele
School of Business
Woxsen University
Hyderabad, Telangana, India

Suhash Pokuri
School of Business
Woxsen University
Hyderabad, Telangana, India

Abhichandra Polkampally
School of Business
Woxsen University
Hyderabad, Telangana, India

V. Kamakshi Prasad
Jawaharlal Nehru Technological University
Hyderabad, Telangana, India

Ruslana Raemgulova
Laboratory of Artificial Intelligence
Skolkovo School of Management
Moscow, Russia

Jeereddy Mani Chenna Kesava Reddy
School of Technology
Woxsen University
Hyderabad, Telangana, India

Vladimir Shabason
SelfOrb
Manama, Bahrain

Simna Shanavas
Presidency College
Bangalore, Karnataka, India

Bhupinder Singh
Sharda University
Greater Noida, Uttar Pradesh, India

Gulam Mahabub Subhani
Department of Computer Science and Engineering
Koneru Lakshmaiah Education Foundation
Hyderabad, Telangana, India

Ch Sumalakshmi
Department of Computer Science and Engineering
Koneru Lakshmaiah Education Foundation
Hyderabad, Telangana, India

Gaikwad Beena Suresh
Presidency College
Bangalore, Karnataka, India

Egor Volkov
Laboratory of Neurosciences in Management
RANEPA
Moscow, Russia

Vladislav Zapylikhin
Artificial Intelligence Laboratory
Skolkovo School of Management
Moscow, Russia

Chapter 1

The Silent Revolution: How AI Is Rewriting the Rules of Organizational Power

Thierry Cellerin

Introduction

The expanding integration of AI in organizational decision-making is reshaping traditional managerial authority and governance structures. Powered by machine learning and predictive analytics, AI-driven systems have the capacity to enhance operational efficiency. At the same time, it alters the mechanisms through which authority is distributed and exercised within firms. Historically, managerial decision-making has relied on hierarchical control, experience-based intuition, and strategic management. However, AI's ability to process vast amounts of data, generate real-time insights, and automate routine cognitive tasks challenges conventional notions of managerial influence and relevance. As AI transforms business processes, organizations will confront a fundamental shift in governance, where power dynamics will no longer be dictated solely by human discretion but mediated by algorithmic-driven decision frameworks.

While existing research has extensively examined AI's impact on efficiency and productivity, a significant theoretical gap exists in understanding how AI redefines power structures within firms. Current literature often

DOI: 10.4324/9781003606956-1

frames AI as a decision-support tool that enhances managerial capabilities but underestimates AI's potential to act as an independent force that reshapes authority relations. Unlike past technological advancements that primarily augmented human decision-making, AI introduces algorithmic governance, a system in which decision authority is embedded within data-driven models rather than hierarchical leadership structures. This chapter seeks to address this gap by exploring how AI influences managerial roles in organizations by redistributing authority and the shifting relevance of expertise.

This question covers a broad set of organizational transformations, including the delegation of decision-making to AI, the evolving role of managerial control, and the restructuring of expertise-based hierarchies. Given the extensiveness and complexity of AI's impact, a comprehensive analysis of these dimensions would be impractical within a single study. To provide a more focused theoretical exploration, this chapter narrows its scope by applying the Power Dynamics Framework (PDF) lens, which emphasizes power's relational and fluid nature. This framework allows us to examine how AI mediates trust, authority, and the shift from traditional managerial roles to algorithmic-driven governance.

By analyzing AI's role through the lens of the PDF, this chapter argues that AI does not simply decentralize decision-making, but instead introduces a new logic of control based on algorithmic governance. Traditional managerial authority, once rooted in positional power and experiential knowledge, is increasingly being transformed toward informational power, where influence is related to the access and interpretation of AI-generated insights. This evolution reconfigures organizational power relations, where managerial control is redistributed across human–AI collaborative systems. As AI systems gain autonomy in shaping business strategies and operational efficiencies, managers must transition from direct decision-makers to meta-governors who oversee AI's ethical and strategic alignment with organizational goals.

The significance of these transformations extends beyond theoretical discourse, as corporations must adapt their leadership models, governance frameworks, and decision-making processes to accommodate AI's growing influence in businesses. This study contributes to academic theory and practical management strategies by offering a nuanced understanding of AI's role as a disruptive force in organizational authority. By conceptualizing AI as an active agent in decision-making, this research highlights the urgent need to structure new governance frameworks that balance AI's analytical capabilities with human administration.

This exploration aims to bridge the gap between technological advancements and traditional power dynamics, shedding light on the challenges and opportunities AI presents in the evolving business landscape. The subsequent sections will further develop this argument by examining AI's impact on managerial authority, expertise relevance, and decision-making structures, ultimately offering a comprehensive contribution to understanding how AI redefines governance in contemporary businesses.

Theoretical Background

As 72% of organizations have already adopted AI in at least one business function as of 2024 [1], these integrations into businesses aim to fundamentally reshape managerial responsibilities. This shift pressures the exclusivity of managerial control in decision-making and the traditional hierarchical structures as a whole. Rather than making decisions themselves, managers might increasingly play a supervisory role in overseeing AI-driven processes and validating algorithmic insights. This evolution could lead to the erosion of managerial authority, as AI's predictive capabilities may eventually surpass human intuition, leading to a dependency on machine-generated recommendations [2]. The paradox of automation and augmentation further compounds this pressure. While AI can enhance managers' decision quality and efficiency, it simultaneously jeopardizes the relevance of their role as a fundamental node in organizational structures. This creates role ambiguity, as managers must balance their potentially diminishing direct decision-making power with the increasing necessity of ensuring that AI-driven decisions align with broader business goals. The result could be a shift toward a more facilitative role, where managers oversee AI-driven decision frameworks rather than serving as the primary decision-makers [3].

AI's integration into decision-making processes disrupts traditional expertise hierarchies, challenging the longstanding link between experience and authority. Traditionally, managerial influence has been closely tied to accumulated industry knowledge, which provided a foundation for strategic decision-making. AI's ability to generate real-time insights and recommendations diminishes the knowledge gap between employees and senior executives, reducing the importance of managerial expertise. Moreover, AI reconfigures organizational learning dynamics by transforming how expertise is acquired and validated. Historical models of knowledge development focus on experiential learning and mentorship, with expertise gained over time. AI, however, enables direct access to expansive knowledge repositories, real-time recommendations, and automated decision-support systems. This accessibility allows even junior employees to engage in decision-making processes with a level of confidence previously reserved for experienced managers, fundamentally altering the pathways through which authority is legitimized [4].

The adoption of AI in organizations is redistributing decision-making power, shifting it from human-exclusive capabilities toward algorithmically influenced recommendations. Previously, authority has been concentrated at the upper managerial levels, with hierarchical structures ensuring a transparent chain of command. AI-powered systems integration disrupts this dynamic by embedding decision-making capabilities into algorithmic processes, leading to a form of authority that is increasingly data-driven rather than experience-based [5]. This transformation forces organizations to redefine their governance models as AI-generated insights and recommendations become central to strategic decision-making. While managers retain a role in overseeing AI applications, their authority is increasingly contingent upon their related ability to command and interpret AI-driven outputs rather than dictate strategic direction themselves. This shift signals a movement toward an algorithmic governance model in which the human complements, rather than dominates, AI-driven decision-making.

AI also accelerates the flattening of organizational hierarchies by reducing reliance on middle management and enabling direct access to decision-relevant insights. Middle managers served traditionally as intermediaries, filtering information between employees and senior leadership. For example, widely available AI-powered employee evaluation systems, predictive analytics modeling, and programmatic media buying tools can bypass these intermediaries, allowing employees at various levels to engage directly with critical business insights [6]. Furthermore, AI automates management functions, shifting managerial roles toward governance of AI-driven systems rather than direct supervision of employees. This transition reinforces the idea of networked decision-making structures, where authority is distributed based on data access rather than hierarchical position. As a result, organizations increasingly resemble decentralized networks rather than rigid hierarchies, fostering a more agile and adaptable business environment [7].

In conclusion, AI integration in corporations fundamentally alters managerial authority, decision-making structures, and expertise relevance. With

AI-driven insights becoming central to decision-making, expertise hierarchies would be redefined as authority is de facto redistributed within organizations. Organizations must, therefore, recalibrate their leadership structures to navigate this evolving landscape, ensuring that AI-driven decision-making frameworks align with strategic goals while maintaining human oversight. The shift from hierarchical control to a networked, data-driven governance model necessitates rethinking leadership competencies, emphasizing AI fluency, ethical supervision, and interpretive judgment. Organizations can fully leverage AI's potential by adopting adaptive governance models while preserving accountability, strategic alignment, and human agency in decision-making.

Theoretical Framework

The Power Dynamics Framework (PDF), introduced by French and Raven [8], refined by Kellerman [9], and further developed in organizational studies, provides a critical lens for analyzing how AI-driven authority shifts reshape decision-making structures. It conceptualizes power as fluid and relational rather than fixed and hierarchical [10]. Therefore, it functions well for analyzing AI's role in organizations. Unlike institutional theory, which primarily focuses on how norms and structures shape decision-making, PDF emphasizes how power continuously shifts based on the actors controlling informational resources. Similarly, while sociotechnical systems theory explores the interplay between technology and human processes, it does not fully capture how AI autonomously alters power relations by embedding authority into decision logic. The advantage of PDF is its ability to explain why AI-driven governance is not simply a shift in decision-making tools but a fundamental transformation of organizational control, where influence is determined by access to AI insights rather than tenure or managerial discretion.

Traditionally, corporate power has been associated with hierarchical control, where managers derive authority from positional status, providing access to critical organizational data and control over resources. AI disrupts this model by embedding decision-making into algorithmic systems, shifting influence away from human intuition to data-driven insights. In doing so, AI fundamentally alters the foundation of authority by shifting power from positional authority, which is rooted in hierarchical structures, to informational and algorithmic power. Therefore, AI democratizes access to strategic insights throughout an organization, reducing the power gap between executives and employees. This shift moves power toward networked decision-making structures, where authority is no longer concentrated in a select few but evenly distributed based on employees' ability to interpret AI-generated insights effectively. The result is a reconfiguration of power relations, where influence is derived from AI fluency rather than hierarchical position.

AI challenges traditional expertise hierarchies by redefining what constitutes authority in decision-making. Before AI integrations, managerial influence originated notably from experience and deep industry knowledge. However, AI's ability to analyze complex datasets in real time reduces the exclusivity of experiential learning, favoring those who can effectively leverage and interpret AI-generated insights. This shift creates new asymmetries in influence as deep industry expertise becomes widely accessible. Consequently, employees with strong AI literacy may exert disproportionate influence over decisions, while traditional expertise relevance may diminish in strategic discussions.

AI fosters, therefore, a collaborative power structure, redistributing decision-making authority between human decision-makers and algorithmic systems. Unlike traditional management models where executives dictated strategy, AI-driven insights increasingly shape operational and strategic decisions. Concretely, if AI-generated recommendations outperform human judgment, it may reduce organizational reliance on managerial intuition. Businesses interested in integrating AI into governance will develop hybrid processes, where AI systems provide analytical guidance while managers oversee contextual validation. This evolution demands a "centerless" organizational structure with adaptive governance as corporations transition from rigid hierarchies to AI-augmented decision ecosystems.

With AI assuming a more prominent role in decision-making, managers transition from direct decision-makers to meta-governors, overseeing AI-driven processes and ensuring ethical compliance [3]. This shift redefines managerial influence, as managers focus less on direct leadership and more on mediating between AI-driven insights and broader organizational objectives. Therefore, integrating AI necessitates new leadership competencies, interpretive judgment, and the ability to navigate ethical dilemmas in algorithmic decision-making. Corporations must prepare managers to balance the efficiency gains of AI with the need for human oversight and accountability. Corporations must cope with AI's predictive capabilities potentially introducing a false sense of objectivity, as algorithms reinforce existing biases, leading to a perceived reality gap between AI-driven and human-driven decision-making. Without a clear governance framework, these organizations risk unintended consequences, where AI appears to flatten hierarchies but instead maintains disproportionate power asymmetries, potentially leading to major operational failures.

Despite its applicability, the PDF theory has limitations when applied to AI governance. The first is the risk of organizations' over-reliance on AI, where becoming excessively dependent on algorithmic decision-making could mislead operations into counter-productivity [11]. While being efficient, AI systems lack contextual awareness and adaptability, potentially leading to flawed strategic choices if left unchecked. Another critical issue is the erosion of institutional knowledge. As AI-generated insights replace traditional expertise, organizations risk losing critical tacit knowledge essential for strategic decision-making. Over-reliance on AI could diminish human problem-solving capabilities and weaken leadership intuition, as employees defer to machine-generated recommendations without critical evaluation. Finally, AI-augmented decision ecosystems raise concerns about AI's ability to adapt to unstructured, novel challenges that require human creativity and contextual judgment.

In conclusion, the literature confirms that organizations lack a structured transition framework to algorithmic governance. AI adoption is often fragmented, with companies implementing AI tools without clear protocols for integrating AI into decision-making processes. This creates ambiguities in accountability, as AI-driven insights can conflict with managerial discretion. Without clear and structured governance models, organizations risk ad-hoc decision-making, where neither AI nor human managers assume full responsibility for critical business outcomes. Addressing this gap requires adaptive leadership frameworks that balance algorithmic decision authority, ethical alignment, and human expertise. The PDF offers a valuable perspective on AI-induced authority shifts, illustrating how AI restructures decision-making, expertise, and managerial roles. AI does not merely flatten hierarchies but redefines governance models, necessitating new forms of power distribution. However, to ensure AI adoption enhances rather than undermines organizational effectiveness, companies must access a framework that balances AI's analytical capabilities with human supervision and proposes transparent accountability mechanisms.

Theoretical Contribution

Artificial intelligence should not be considered a new decision-making enhancement tool, but an autonomous force that actively reshapes organizational power structures. Unlike previous technological advancements that enhanced managerial efficiency, such as personal computers and the internet, AI systems increasingly integrate decisional functions, influencing governance beyond human control. This transformation challenges traditional leadership paradigms, as AI-driven insights shape operational priorities, often superseding managerial intuition and experience [12].

Organizations require new leadership frameworks to accommodate AI's expanding role. Firms must adopt hybrid leadership models, integrating AI into operational workflows while maintaining human interpretive authority. This shift will necessitate revised governance structures that define the boundaries of AI autonomy, ensuring that machine-driven decisions remain aligned with corporate values and ethical considerations. It also involves enhancing AI literacy among managers, which will be crucial in maintaining a balanced governance model. By integrating AI capabilities with human oversight, businesses can foster a more adaptive and resilient decision-making structure that leverages AI's strengths while mitigating potential risks. A hybrid governance approach would allow managers to retain control while using AI-generated insights to support decision-making rather than fully automating it. For example, some firms establish AI supervisory boards to evaluate algorithmic decisions, detect biases, and ensure compliance, particularly in regulated industries. Another approach would involve tiered decision frameworks, where AI handles low-risk tasks autonomously, moderate-risk decisions involve human–AI collaboration, and high-risk decisions remain under human authority. These models could help organizations harness AI's potential while maintaining human supervision in critical areas, ensuring AI is a tool for enhanced decision-making rather than an unchecked authority.

The evolving role of AI necessitates a fundamental shift in managerial competencies. Traditional leadership models emphasizing directive orders from managers to subordinates are disrupted through AI-augmented governance, where managers act as facilitators rather than sole decision-makers. To navigate this transition, managers must develop data-fluent mindsets, ensuring they can critically interpret algorithmic outputs and ensure that these align with organizational goals [13]. Data-fluent managers will be better equipped to balance algorithmic decision-making with contextual judgment, preserving human oversight where necessary.

The theoretical implications of AI's role as a decisional actor are significant. Current research predominantly frames AI as a neutral enabler of efficiency, neglecting its potential to restructure authority dynamics and assume managerial responsibility. AI-driven decision frameworks introduce a new locus of power, where authority is no longer derived solely from hierarchical position but from algorithmic influence and data control. As algorithmic governance expands, questions about who ultimately controls decision-making between human executives and algorithmic intelligence will arise. Organizations must address this shift to prevent power imbalances from undermining managerial autonomy, leading to potential operational disruptions.

In conclusion, AI's growing influence in organizational decision-making necessitates rethinking theoretical and managerial perspectives. Recognizing AI as a decisional actor highlights the need for a structured framework to navigate shifting power dynamics while enhancing AI literacy among managers, which will be crucial in maintaining a balanced governance model.

Conclusion

This study explores the notion that AI is increasingly acting as a force that actively reshapes managerial authority, governance structures, and expertise hierarchies. Through the PDF, this research demonstrates how AI shifts power from positional authority traditionally held by managers to informational and algorithmic governance. This transition will fundamentally alter how decisions are made, challenging longstanding hierarchical structures and promoting networked, AI-augmented decision ecosystems. Managers will increasingly evolve from direct decision-makers to meta-governors in charge of aligning AI-driven insights with corporate goals. The power structure within organizations could become more fluid and relational rather than static and top-down. AI does not simply decentralize decision-making; it reconfigures governance by embedding decision logic into automated algorithmic systems. However, while this shift enables greater synergetic efficiency, it also introduces significant governance challenges, particularly regarding accountability, transparency, and ethical alignment.

If the competitiveness gains promised by AI for companies are evident, the potential over-reliance on algorithmic systems and the risk of losing institutional knowledge are significant concerns that organizations must address. A simple case would be potential friction between algorithmic decisions and human judgment or organizational values. The PDF provides a valuable lens for understanding these shifts. However, it also highlights the need for adaptive governance models that balance AI's analytical capabilities with human supervision, ethical and juridical considerations, as well as clear accountability mechanisms. The insights highlighted in this study open several directions for future research. A first interrogation would concern how different organizational contexts – such as industry type, company size, and cultural factors – could affect the adoption and impact of AI on power dynamics. Regulated industries like healthcare or finance could have strong exigencies against AI-driven decision-making. Another topic of interest could be related to the long-term effects of AI on organizational knowledge retention. As AI systems increasingly replace traditional expertise, how can companies ensure that the critical tacit knowledge source of added value is preserved and integrated into AI-driven processes? Finally, more studies could explore the ethical implications of AI-driven governance as AI systems gain more autonomy in decision-making.

Corporations need clear guidelines on framing AI decision authority, ethical responsibility, and managerial accountability to transition toward algorithmic governance. Today's academic literature lacks dedicated frameworks that could operationalize context-specific governance strategies tailored to different industries, regulatory environments, or organizational cultures. Future research should define such frameworks, offering guidance for businesses to harness the true potential of AI while preserving human oversight, ethical accountability, and institutional knowledge. The journey toward AI-augmented governance is only beginning, and the lessons learned today will profoundly shape the future of all organizations.

References

1. Statista. (2024). *Adoption of artificial intelligence among organizations worldwide from 2017 to 2024, by type*. Retrieved February 4, 2025, from https://www.statista.com/statistics/1545783/ai-adoption-among-organizations-worldwide/
2. Von Krogh, G. (2018). Artificial intelligence in organizations: New opportunities for phenomenon-based theorizing. *Academy of Management Discoveries*, *4*(4), 404–409. https://doi.org/10.5465/amd.2018.0084
3. Raisch, S., & Krakowski, S. (2021). Artificial intelligence and management: The automation–augmentation paradox. *Academy of Management Review*, *46*(1), 192–210. https://doi.org/10.5465/amr.2018.0072
4. Shrestha, Y. R., Krishna, V., & Von Krogh, G. (2021). Augmenting organizational decision-making with deep learning algorithms: Principles, promises, and challenges. *Journal of Business Research*, *123*, 588–603.
5. Jarrahi, M. H., Newlands, G., Lee, M. K., Wolf, C. T., Kinder, E., & Sutherland, W. (2021). Algorithmic management in a work context. *Big Data & Society*, *8*(2), 20539517211020332. https://doi.org/10.1177/20539517211020332
6. Duan, Y., Edwards, J. S., & Dwivedi, Y. K. (2019). Artificial intelligence for decision making in the era of Big Data – evolution, challenges and research agenda. *International Journal of Information Management*, *48*, 63–71.
7. Fountaine, T., McCarthy, B., & Saleh, T. (2019). Building the AI-powered organization. *Harvard Business Review*, *97*(4), 67–73.
8. French, J. R. P., & Raven, B. (1959). *The bases of social power*. Institute for Social Research.
9. Kellerman, B. (Ed.). (1986). *Political leadership: A source book*. University of Pittsburgh Press. https://doi.org/10.2307/jj.12381759
10. Clegg, S. (1989). *Frameworks of power*. Sage Publications. https://www.torrossa.com/en/resources/an/5543042
11. Solberg, E., Kaarstad, M., Eitrheim, M. H. R., Bisio, R., Reegård, K., & Bloch, M. (2022). A conceptual model of trust, perceived risk, and reliance on AI decision aids. *Group & Organization Management*, *47*(2), 187–222.
12. Sowa, K., Przegalinska, A., & Ciechanowski, L. (2021). Cobots in knowledge work. *Journal of Business Research*, *125*, 135–142. https://doi.org/10.1016/j.jbusres.2020.11.038
13. Barrotta, P., & Montuschi, E. (2018). Expertise, relevance and types of knowledge. *Social Epistemology*, *32*(6), 387–396. https://doi.org/10.1080/02691728.2018.1546345

Chapter 2

Equitable AI: Aligning Human Expertise with Algorithmic Value

Evgeny Goryunov and Alexander Didenko

Introduction

Data annotators, the humans who label text, images, audio, and video to train AI models, form the invisible backbone of the AI industry. Labeling tasks performed by skilled human workers (ranging from semantic segmentation of medical images to the transcription of complex financial documents) often involve nuanced judgment and domain-specific expertise. Research indicates that errors in data annotation are responsible for up to 42% failures in the industrial deployment of AI solutions. Furthermore, as will be demonstrated later, employees might have a rational motive to intentionally distort data if they are concerned that sharing tacit knowledge with AI could undermine their human capital.

Tasks, Pay, and Key Concerns

Data annotation needs to span all major industries. Table 2.1 summarizes typical annotation tasks, average compensation, and key labor concerns in each sector.

DOI: 10.4324/9781003606956-2

Table 2.1 Overview of Data Annotation Tasks, Typical Compensation, and Labor Concerns of Industry (2022–2025)

Industry/Sector	*Typical Annotation Tasks*	*Average Pay and Fees*	*Key Concerns*
Technology and social media (IT, communication services)	– Content moderation (flagging hate, violence, etc.). – Search relevance rating. – Chatbot training (intent labeling, conversations). – Image tagging for AR/social apps.	Often outsourced to global contractors at \$1–\$3/h (e.g., content moderators in Kenya for OpenAI reportedly <\$2/h). In high-cost countries, \$15/h for in-house roles (e.g., US-based content moderators).	– Low pay and wage theft. – Traumatic content exposure. – Misclassification of workers as contractors (e.g., Sama/Meta case).
Automotive (autonomous vehicles)	– Image/video annotation (bounding boxes for pedestrians, vehicles, road signs). – LiDAR/point cloud labeling. – Driving scenario tagging.	Often crowdsourced or outsourced (\$1–\$5/h in Global South data farms). Experienced annotators in China: 4,000+ CNY/month (\$560).	– Repetitive, high-volume tasks. – Low wages ("race to the bottom" in global bidding). – Exploitation of student interns in China.
Healthcare and pharmaceuticals	– Medical image annotation (X-rays, MRIs, pathology slides). – Transcription and tagging of doctor–patient conversations. – Genomic data labeling.	Higher-skilled domain experts required. \$10–\$50/h for physicians (project-based). Low-end transcriptions outsourced at \$5–\$15/h.	– Shortage of qualified annotators. – Exploitation of med students in lower-income countries. – Quality concerns impacting patient safety.

(*Continued*)

Table 2.1 (Continued)

Finance and banking	– Document classification (invoices, loans). – Fraud detection (transaction labeling). – Sentiment annotation (market-related news).	Often outsourced to BPO firms (India, Philippines). \$3–\$6/h for basic tagging. Specialized financial annotation: \$10–\$20/h.	– Accuracy critical (errors propagate into financial models). – Confidentiality issues. – High-volume work leads to burnout.
Retail and e-commerce	– Product tagging and categorization. – Recommendation data annotation. – Customer review sentiment analysis.	Crowdwork common (Amazon Mechanical Turk). \$3/h median pay, often piecewise. In-house annotators earn \$12–\$18/h in high-cost regions.	– Gig work with no benefits. – Task rejection without pay. – Global competition drives wages down.
Media and entertainment	– Audio/video transcription and tagging. – Metadata labeling (genres, themes). – Moderation of user content.	Mostly outsourced. \$2–\$4/hour in low-cost regions. Skilled captioners (bilingual): \$10/h.	– Tight deadlines. – Unpaid "volunteer" labor in anime/ gaming industries. – Cultural misinterpretations in tagging.
Government and security	– Surveillance image labeling. – Open-source intelligence annotation. – Document redaction/ classification.	Contracted via defense/security firms. Some projects pay a few cents per image, others require clearance (\$15–\$25/h).	– Ethical concerns (workers unaware of surveillance impact). – Exposure to graphic content. – Lack of job security despite sensitive work.

Important to note that across all sectors, annotation tasks range from simple repetitive clicks to specialized judgment calls. However, a common theme is the outsourcing of labor to lower-wage regions and the classification of workers as independent contractors, leading to stark inequalities between the value generated for AI companies and the compensation received by annotators.

Controversies and Exploitation

Several high-profile cases and investigations in recent years have spotlighted the harsh realities faced by data laborers [1].

- **OpenAI and Kenyan Content Moderators (2021–2023):** In a bid to filter toxic content for ChatGPT, OpenAI contracted Sama (an outsourcing firm in Nairobi). A 2023 *Time* investigation revealed workers were paid as little as $1.32–$2 per hour to read and label extremely disturbing texts and images (hate speech, violent, sexual content). Many moderators reported psychological trauma from the exposure. After the project, nearly 200 Kenyan workers were abruptly laid off in 2022. This "mental trauma for a few dollars" scandal led to a public outcry and raised questions about Big Tech's responsibility to its invisible workers [2].
- **Meta (Facebook) and Sama Lawsuit in Kenya (2022–ongoing):** Sama was also the main subcontractor for Facebook's content moderation in Africa. In 2022, whistleblower Daniel Motaung sued Meta and Sama in a Nairobi court, alleging unreasonable working conditions, union-busting, and even "human trafficking"-like recruitment practices. Moderators for Facebook in Kenya described irregular pay, inadequate mental health support, and being prevented from organizing. Meta claimed the workers weren't its employees, but the Kenyan courts have thus far allowed the case to proceed [3]. This case is precedent-setting – it's one of the first times a Big Tech firm faces legal accountability in an African court for labor practices. (Notably, Meta had previously paid $85 million to settle a US class-action by content moderators over PTSD.)
- **Scale AI's "Digital Sweatshops":** Scale AI, a US-based unicorn startup valued at ~$7 billion, runs the Remotasks platform with over 10,000 workers in the Philippines. Investigations by *The Independent* (UK) found systematic exploitation: extremely low rates per task and frequent delays or non-payment of earnings. Workers ("taskers") reported that after completing work, payments would be withheld or

canceled with little recourse – some saw earnings far below the local minimum wage (Philippine daily min wage ~$6–$10, but taskers often earned less). Around 2021, as Remotasks expanded globally, pay rates plunged – tasks that once paid $10 each fell to mere cents. This effectively created a "race to the bottom" in wages, pitting workers in different countries against each other for ever-smaller payments. Labor advocates labeled these conditions "digital sweatshops," noting the absence of regulation for this informal gig work. Scale AI claims to be improving its pay system, but this case highlights how AI's rapid growth has been built on cheap, disempowered labor [4].

- **CrowdFlower (now Figure Eight/Appen) Lawsuit (2015):** One of the earliest legal challenges in the crowd work arena was a class-action against CrowdFlower, a platform posting micro-tasks. Workers alleged they were misclassified as contractors and paid below minimum wage – sometimes as low as $2/h. In 2015, the company settled for $585,000 – a modest sum but a symbolic victory. It underscored the gig economy issue of wage theft (paying piecemeal rates that translate to sub-minimum wages) and set a precedent that such contributors could claim labor rights. However, even after settlement, no guarantee was made to pay minimum wage going forward on all platforms, and crowd work platforms proliferated with similar models [5].
- **Exploitation of Chinese Student Interns (2023):** A *Rest of World* investigation found that data labeling companies in China partner with vocational schools to recruit student interns as cheap labor. Thousands of students have been required to do data annotation (tagging images, filtering content, etc.) as part of their graduation requirements, often for sub-minimum wages and long hours. These interns, essentially captive workers, bolster AI projects for tech giants (Alibaba, Baidu, etc.) while enduring poor conditions that raise human rights concerns. The practice saves companies money but blurs the line between education and labor exploitation [6].

These case studies reveal common themes: outsourcing to vulnerable populations, pay far out of proportion to the value of the work, suppression of worker voices, and in some instances severe psychological toll (especially in content moderation scenarios dealing with horrific material). They have sparked calls for better oversight – for example, after the Kenyan content moderation revelations, there are growing demands for Big Tech to adhere to global labor standards even when operating in developing countries.

Regional Disparities in Pay and Conditions

There are stark regional disparities in how data annotators are compensated and treated, often reflecting global economic inequalities:

- **Global North vs. Global South:** The highest concentration of data annotation work is in the Global South (Asia, Africa, Latin America), where English-speaking educated workers are abundant and local wages are low. India alone accounts for about one-third of global online freelancers (many doing data labeling), and developing countries together account for ~66% of the remote gig workforce. This workforce is paid a fraction of what similar work might earn in the US or Europe. For example, Indian crowdworkers on Amazon MTurk often make only $10–$30 per day after many hours of work – which can equate to roughly $1–$3/h given the high rejection rates on tasks. By contrast, an in-house annotator in the US (e.g., at a tech company's office) might make $15–$20/h plus benefits. However, such roles in the Global North are limited; the bulk of annotation is offshored [6].
- **Asia (India, Philippines, etc.):** South and Southeast Asia have become major hubs. In the Philippines, with its young, English-speaking workforce, an estimated 2 million people engage in online "crowdwork" including AI data tagging. Reported earnings for Filipino annotators on platforms like Remotasks often fall below the legal minimum wage, e.g., just a few dollars a day for hours of work [7]. In India, competition on crowdsourcing platforms is fierce; workers may stay up overnight to grab US tasks. Many Indian annotators treat it as piecemeal supplementary income due to low pay, but for some it's a primary job despite the lack of stability and meager pay [8]. Both regions suffer from lax labor regulation for freelancers – laws haven't caught up to protect gig workers, enabling what one expert calls "precarious ... bottom of the AI value chain" work.
- **Africa:** African countries (Kenya, Nigeria, Ethiopia, etc.) have also tapped into the data labeling boom. Kenya has become an outsourcing center for content moderation and transcription (due to high English proficiency). Kenyan AI annotation workers often earn around $1–$3 per hour, as seen in the Sama case [9]. While this can be above Kenya's official minimum wage, the cost in mental health and the disparity with Big Tech profits have been heavily criticized. In Nigeria, annotators working for overseas clients (e.g., a U.S. firm "Hugo" hiring AI raters)

report being monitored to the second and not paid for overtime, essentially squeezed for maximum output within an 8-hour shift [10]. Africa's annotation workforce often faces unreliable pay (delays, non-payment) and little to no labor protections if problems arise, given weak enforcement locally and the difficulty of holding a foreign company accountable.

- **China:** Within China, the data annotation industry has grown domestically to fuel Chinese AI firms. Pay in inland Chinese cities for annotators (often young college graduates or migrant workers) might be on the order of ¥3,000–¥4,000 RMB/month ($400–$560), which, while modest, can be a stable income for those regions. New hires or interns may start as low as ¥1,000–¥2,000/month (only $140–$280). Labor conditions in China can be harsh – reports describe "assembly line" workloads and physical exhaustion. The use of student interns for cheap labor (as noted above) is a distinct issue in China, effectively exploiting an even lower-paid tier. On the other hand, the Chinese government is actively investing in data labeling bases in various cities, indicating an official push to structure and possibly standardize this workforce. It remains to be seen if that improves wages or just output.
- **United States and Europe:** In high-income countries, relatively few people work full-time on basic data annotation – those that do are either employed in specialized roles or participate via platforms as a side gig. Crowdsourced workers in the US on MTurk also earn very little (median around $3/h, well below the US minimum wage) [11]. Some states in the US have started to examine whether crowdworkers should be treated as employees; the CrowdFlower case in California (though settled) brought attention to this [12]. In Europe, the EU is discussing regulations for platform workers, which could improve conditions if annotation falls under its scope. For now, companies often route tasks away from high-wage Western markets. One exception is content moderation: due to language and cultural needs, some moderation and labeling jobs exist in the West (for example, Dublin, Ireland, hosts Facebook's content moderation hub for Europe). Those moderators might earn around €12–€14/h and have labor law protections, yet still reported high stress and felt underpaid for the emotional toll [13]. The disparity even within the same company is telling: a Facebook moderator in Ireland could be making five to six times the wage of a moderator in the Philippines – highlighting global inequities for essentially the same work [14].

Impact of Underpayment on AI Accuracy and Bias

The practice of underpaying and undervaluing data annotators doesn't only raise moral and legal issues – it can also directly affect the performance and fairness of AI systems:

- **Data Quality and Accuracy:** There is a well-known maxim: "Garbage in, garbage out." If annotators are pressed to work faster for unsustainably low pay, the quality of annotations can suffer. Workers who are juggling dozens of tasks an hour to make ends meet are likely to make mistakes or apply shallow judgment. Indeed, research finds that better-paid annotators tend to produce higher-quality labels – one study noted higher pay correlates with more accurate performance on tasks [8]. Conversely, inconsistent or erroneous labels (for example, mislabeled images or incorrect transcript annotations) will degrade an AI model's accuracy since the training data is noisy. Low-quality data yields low-quality AI. Companies that prioritize cutting annotation costs may actually incur hidden costs later when models trained on sloppy labels perform poorly [15].
- **Skewed Sample of Annotators:** Underpaying means only certain demographics (those in low-income regions, or those desperate for any income) will take the job. This lack of diversity in who annotates data can introduce bias. For example, if most image annotators come from a specific country or background, their interpretation of images or cultural symbols could bias the dataset. Annotators might unintentionally label things through the lens of their own context (e.g., identifying an object incorrectly because it's unfamiliar in their region) [16]. In natural language processing (NLP) tasks, the understanding of slang or context could differ. A more diverse, well-compensated annotator pool would likely yield more balanced data. Underpayment works against that by excluding qualified experts (who won't work for pennies) and concentrating work in certain locales [17].
- **Loss of Domain Expertise:** Some AI applications require domain knowledge – e.g., medical or legal data annotation, as noted. If budgets are too low, companies might forego using actual experts and instead use less-trained annotators, increasing the chance of mistakes. In medical AI, for instance, if annotators are not sufficiently trained or are rushing because they're underpaid, the labels (say, on X-ray images) may be incorrect and lead the AI to learn wrong patterns. This can cause

models to have lower accuracy or dangerous blind spots. Fair compensation can attract professionals who ensure labels are correct and nuanced [9].

- **Ethical and Bias Issues in Content Moderation:** Content moderation data is used to train AI filters (for hate speech, misinformation, etc.). If moderators are traumatized or burnt out due to harsh conditions, they may become desensitized or overly cautious, affecting labeling consistency. Additionally, if underpaid moderators quit frequently (high turnover), new untrained workers might label differently than predecessors, introducing inconsistency. Bias can also creep in if workers from certain cultural contexts label certain speech as hateful or not, differently from other cultures [18]. Without investment in training and fair pay, these differences aren't ironed out, and the AI may inherit those biases, impacting its fairness across demographics.
- **Reduced Accountability in AI Outputs:** Underpaid annotators often have no channel to provide feedback or question the labeling guidelines – they're expected to follow orders quickly [13]. This can be problematic if the guidelines themselves are biased or flawed. A well-supported annotator workforce could flag issues in data or labeling policies (e.g., noticing that certain genders or ethnicities are being consistently labeled negatively, etc.). But an exploited workforce likely stays silent, either out of fear or lack of time, meaning potential biases in datasets go unreported [11, 16]. The result is AI systems that mirror societal biases or the blind spots of overstressed workers.
- **Sustainability of AI Development:** In the long run, if the controversies around exploitation continue, there may be labor shortages or stricter regulations that catch companies off guard. Already, some African and Asian workers are organizing (such as the Kenyan moderators' petition and lawsuits) and calling for better pay [12, 14, 15]. If annotation remains a dead-end "gig" with poor pay, turnover will remain high, and experienced annotators will be rare. This continual loss of experienced laborers can hurt AI projects – models might take longer to reach high accuracy because new workers make beginner mistakes. Thus, underpayment can slow down the iterative improvement of AI models [19].

Mitigating the Impact

There is growing recognition that fair compensation and training for annotators is not just a moral imperative but also crucial for AI quality. Some companies and researchers are experimenting with approaches like paying above minimum wage globally (Appen claims it pays above local minimums to its contributors [16]) or implementing quality bonus systems. The gig platforms that succeeded by driving costs down are now facing backlash and legal risks that may force change [12, 15, 16, 19]. Ultimately, investing in the "human in the loop" leads to better AI outcomes, because well-paid, motivated annotators can deliver the careful, unbiased work that cutting-edge AI development demands [8, 9].

The last few years have laid bare the inequalities in the AI supply chain. Data annotators worldwide perform the meticulous labor that allows AI models to be accurate and reliable, yet their compensation often pales in comparison to the value they generate – and in many cases, falls below decent work standards. From the content moderators in Nairobi filtering the darkest corners of the internet for $2/h [10–12], to the click-farm crowdworkers in Manila and Mumbai racing to label e-commerce products [20], to the underpaid interns in China's AI factories [21], these workers share common struggles: low wages, tenuous employment, and sometimes harmful working conditions. The controversies and lawsuits emerging globally (Kenya, California, Philippines, etc.) indicate a turning point [12, 14–16, 19].

Ensuring fair pay and safe working conditions for annotators is not only a matter of justice but also integral to building trustworthy AI systems. As the industry matures, stakeholders – AI companies, clients, regulators, and consumers – are beginning to ask who is behind the "AI magic." The answer is millions of human workers, and recognizing their contributions through better compensation and rights will be key to sustaining AI's growth. High-quality AI is built on high-quality data, and high-quality data comes from valuing the people who create it [8, 9].

Literature Survey

Data attribution and reward mechanisms have become central topics in data-centric machine learning research, particularly when the goal is to fairly compensate human annotators for the utility their labels add to model performance. The most productive and recent line of research in this area has predominantly relied on cooperative game theory concepts, most notably the Shapley value. This approach both quantifies the marginal contribution of individual data points, thus contributing value to data science workflow, and helps to inform incentive-compatible reward distribution [2, 6]. At the same time, complementary or, sometimes, alternative approaches based on economic and behavioral models were developed that provide alternative frameworks that more directly integrate human factors into incentive design [5].

In general, the primary approaches include:

1. **Leave-One-Out (LOO) Method:** This method evaluates the importance of a data point by measuring the model's performance decrease when that point is removed. While simple, it is computationally expensive for large datasets. The methodology is characterized by its ease of comprehension and execution; however, it incurs a high computational cost, particularly when applied to extensive datasets.
2. **Leverage Scores:** These scores assess the influence of data points based on how much they affect the model's parameters. They are efficient but may not capture all aspects of a data point's contribution. This method demonstrates a high degree of efficiency and offers profound insights into the influence exerted by various data elements. On the other hand, this approach may neglect specific contributions and exhibits a lesser degree of equity when compared to the Shapley values.
3. **Pure Data Shapley Methods:** These methods provide a principled, axiomatic framework for valuing data through properties such as fairness, symmetry, additivity, and the null player condition [2]. Within this framework, each data point is assigned a contribution score that reflects its incremental impact on a given utility measure – typically a downstream model's accuracy. However, the computational burden of determining the exact Shapley value is formidable because it traditionally requires retraining models over the entire power set of the data. To address this, researchers have introduced efficient sampling techniques and approximation methods based on gradient computations, Taylor expansions, and in-run attribution approaches that significantly reduce computational overhead [3, 6].
   a. **Accelerated Shapley Approximations:** Techniques such as ε-Shapley and CHG (Conduct of Hardness and Gradient) Shapley use

gradient-based approximations and Taylor expansions that leverage structural properties of the learning algorithm, reducing the computational burden by up to an order of magnitude [3].

b. **Efficient Sampling:** Methods such as Monte Carlo integration and discrete uniform Shapley reduce computational cost by sampling from the power set, while maintaining robust approximation guarantees [6].

c. **Closed-form and In-Run Approaches:** In scenarios like foundation model pretraining, In-Run Data Shapley circumvents multiple retraining loops by integrating attribution directly into the model's training process [6].

4. **Incentive Structures via Shapley-Based Valuation:** Complementary to these technical improvements are incentive schemes based on the same data attribution principles. Reward mechanisms can be structured such that a data point's higher Shapley value justifies a higher reward, thereby aligning annotator compensation with the tangible impact of their contributions [5]. Furthermore, economic models – such as those derived from principal–agent theory – incorporate annotator heterogeneity, effort costs, and intrinsic motivational factors to design bonus schemes or contract structures that support both annotators and data consumers [1].
5. **Economic and Behavioral Models:** Beyond the strictly quantitative data valuation approaches, research on crowdsourced annotation has highlighted the importance of coupling clear, rule-based task instructions with reliable, modest monetary incentives. Studies have demonstrated that such an approach leads to better annotation accuracy and higher worker satisfaction compared to systems that rely solely on high but uncertain financial rewards [4]. For instance, formulations derived from continuous action spaces reveal that both binary and linear contracts can be optimized by analyzing the gap between first-best and second-best solution regimes. Thus, effective annotation reward systems must combine data-centric utility estimation with worker-centric behavioral insights to achieve both efficiency and fairness [1].
6. **Hybrid Mechanisms and Weak Supervision:** In contexts where annotations are derived from weak supervision sources (for example, programmatically generated labels or LLM-based annotations), methods such as WeShap have been proposed [7]. These rely on a Shapley-driven evaluation to identify beneficial versus detrimental sources among multiple weak annotators. Although these methods are not directly

rewarding human annotators, insights drawn from such frameworks can also be extended to reward systems for the latter [7].

Baseline Model

In our model, there are two agents, Mario and Luigi, who perform tasks. In this version of the model, there are only two possible types of tasks: A and B. We assume that the agents incur no costs in executing the tasks and that they create an added value, which depends both on the type of task and on the agent performing it. Let the added values generated by Mario when solving tasks of type A and type B be denoted, respectively, by X_M^A and X_M^B. Analogously, we denote the corresponding added values produced by Luigi as X_L^A and X_L^B and we assume that Mario is more efficient at performing tasks of type A, whereas Luigi is better suited for tasks of type B. This implies that

$$X_M^A > X_L^A \text{ and } X_M^B < X_L^B$$

At each point in time, exactly two tasks arrive, and any combination of task types is possible. With probability q_{AA}, both tasks are of type A; with probability q_{BB}, both tasks are of type B; and with probability q_{AB}, one task is of type A and the other is of type B.

Once the types of tasks become known, Mario and Luigi choose which task to respond to. We assume that each of them may respond to exactly one task and that neither agent knows to which task the other has responded. After the clients' posting, the tasks receive responses from the agents, each client decides which agent to hire for executing the task. If a client receives only one response, the task is assigned to that agent. If a client receives two responses, the client selects the agent who is more efficient in performing the task. That is, in the case where both agents respond to a given task, the client will assign the task to Mario if it is of type A and to Luigi if it is of type B. We assume that, in negotiations with the client regarding the size of the remuneration for task execution, each agent possesses full bargaining power; therefore, the payment for executing the task is equal to the corresponding added value, i.e., if agent i performs a task of type t, he receives a payment of X_i^t.

Under the assumptions described above, the following scenarios are possible:

a. Both tasks are of type A: Since neither Mario nor Luigi has any preference, each agent randomly and equally likely selects a task to respond to. With a probability of 50%, both Mario and Luigi will submit a response to the same task, leaving the other task without any response. In this case, the contest is won by Mario – being more efficient at performing task A – and he receives a payoff of X_M^A, while Luigi receives nothing (since we assume he cannot take on the other task after being rejected). With the remaining probability of 50%, Mario and Luigi will submit responses to tasks posted by different clients, and then each will perform his respective task, whereby Mario receives a payment of X_M^A and Luigi receives X_L^A. Thus, in the first scenario, Mario always obtains X_M^A, while Luigi's average payoff is only $\frac{X_L^A}{2}$.
b. Both tasks are of type B: The situation is analogous, but in this case Luigi holds the advantage and ultimately receives X_L^B, while Mario, on average, earns $\frac{X_M^B}{2}$.
c. Tasks are of different types (AB): The agents respond in accordance with their respective advantages: Mario responds to the task of type A and Luigi to the task of type B. Then, each of them performs the task, receiving, respectively, X_M^A and X_L^B.

The average payoffs of the agents for each of the possible outcomes are presented in Table 2.2.

ML Corporation

Now suppose that a corporation is established that manufactures robots capable of performing tasks as well. The production of robots incurs negligible costs; hence, we assume that the corporation can respond to several orders simultaneously, thereby being able to reap gains due to economies of scale. However, in order for the robots to be capable of executing tasks, they must be trained, and for that, it is necessary first to engage the agents.

The corporation thus becomes a natural competitor to Mario and Luigi, since the robots are potentially capable of leaving the agents without orders,

Table 2.2 The Average Payoffs of the Agents

	Type AA	*Type BB*	*Type AB*
Mario	X_M^A	$\frac{X_M^B}{2}$	X_M^A
Luigi	$\frac{X_L^A}{2}$	X_L^B	X_L^B

completely displacing them from the market and depriving them of income. Moreover, we assume that, provided a robot and an agent have identical productivity, the client will always prefer to hire the robot; that is, if the robots are trained to perform tasks at least as well as the agents, they will completely oust the agents from the market. For this reason, the agents will demand significant compensation from the corporation if it engages them for training, as their services will subsequently become redundant.

Assume that the corporation offers the agents the following contract: in exchange for undertaking the training of the corporation's robots, they will receive payments of W_M for Mario and W_L for Luigi. The values W_M and W_L are equal to the agents' original average payoffs with an added premium, ϕ_M for Mario and ϕ_L for Luigi. The compensations ϕ_M and ϕ_L are determined by the Shapley values for the cooperative game in which the coalition is defined as the set of agents who agree to accept the corporation's proposal, and the coalition's payoff is the total expected added value that the corporation can generate, taking into account which agents have agreed to participate in the training of the robots.

To calculate the values of ϕ_M and ϕ_L, it is necessary to analyze the payments that the agents would receive depending on which of them accept the offer and share their expertise with the corporation. Let the set of coalitions be denoted by $\mathbf{S}$ and its elements by S. Consider the possible coalition variants (Table 2.3).

Synergy Effects

Let us denote by the column vector

$$q = \begin{pmatrix} q_{AA} \\ q_{BB} \\ q_{AB} \end{pmatrix}$$

Table 2.3 Corporation's Payoff Vectors by Coalition

Coalition	*AA*	*BB*	*AB*
$S = \Omega(\pi_0)$	0	0	0
$S = \{M\}(\pi_M)$	$2X_M^A$	X_M^A	X_M^A
$S = \{L\}(\pi_L)$	X_L^A	$2X_L^B$	X_L^B
$S = \{M,L\}(\pi_{ML})$	$2X_M^A$	$2X_L^B$	$X_M^A + X_L^B$

the probabilities associated with the arrival of different combinations of task requests. Then, the average payoffs to the corporation, contingent on which agents are engaged in training the robots, can be expressed as follows:

$$V_0 = q'\pi_0, V_M = q'\pi_M, V_L = q'\pi_L, V_{ML} = q'\pi ML,$$

The expected payoff for Mario and Luigi in the baseline scenario is given by:

$$W_M = (q_{AA} + q_{AB})X_M^A + q_{BB}\frac{X_M^B}{2}$$

$$W_L = (q_{BB} + q_{AB})X_L^B + q_{AA}\frac{X_L^A}{2}$$

By definition, the Shapley values for Mario and Luigi are:

$$\phi_M = \frac{1}{2}(V_M - V_0) + \frac{1}{2}(V_{ML} - V_L)$$

$$\phi_L = \frac{1}{2}(V_L - V_0) + \frac{1}{2}(V_{ML} - V_M)$$

It follows that the sum of these values, i.e., the total premium paid by the corporation to the agents, is given by:

$$\phi_M + \phi_L = V_{ML} - V_L$$

Denoting this quantity by

$$V_{ML}^{¿} = v_{ML} - v_0$$

we can express the total payments to each agent as:

$$W_M = w_M + V_{ML}^{¿} + \theta,$$

$$W_L = w_L + \frac{V_{ML}^{¿}}{2} - \theta,$$

where the parameter θ is defined by

$$\theta = \frac{V_{M-}V_{L}}{2}$$

Intuitively, θ represents the difference between the payoff the corporation would obtain if it engaged only Mario and the payoff it would receive if it engaged only Luigi.

Now, consider the case in which the engagement of both agents results in a synergistic enhancement of robot productivity. Suppose that if the corporation engages both agents for training, then each robot is capable of generating an added value of X_{syn}^{A} when performing task A and X_{syn}^{B} when performing task B. Assume that

$$X_{syn}^{A} > X_{M}^{A} > X_{L}^{A} \text{ and } X_{syn}^{B} > X_{L}^{B} > X_{M}^{B}$$

Then, the corporation's payoff vector becomes:

$$\pi_{ML}^{syn} = \begin{pmatrix} 2X_{syn}^{A} \\ 2X_{syn}^{B} \\ X_{syn}^{A} + X_{syn}^{B} \end{pmatrix}$$

and the corresponding corporation's payoff is:

$$V_{ML}^{syn} = q'\pi_{ML}^{syn}$$

Under this synergistic scenario, the Shapley values are:

$$\phi_M = \frac{1}{2}(V_M - V_0) + \frac{1}{2}(V_{ML}^{syn} - V_L),$$

$$\phi_L = \frac{1}{2}(V_L - V_0) + \frac{1}{2}(V_{ML}^{syn} - V_M)$$

We propose the following decomposition for the agents' total payments:

$$W_M = W_M + \frac{V_{ML}^{syn}}{2} + \frac{V_{ML}^{scl}}{2} + \theta,$$

$$W_L = W_L + \frac{V_{ML}^{syn}}{2} + \frac{V_{ML}^{scl}}{2} - \theta.$$

Here, V_{ML}^{syn} denotes the additional value generated by the synergistic improvement in robot productivity, and V_{ML}^{scl} represents the additional value derived from the scaling effect.

Enhancement of Robot Productivity

In the previous model, the agents were symmetric. They did not engage in direct competition with one another because each possessed a comparative advantage. As a consequence, each agent maintained its own strengths. In Model 2, however, we consider a scenario in which the joint involvement of both agents in training the corporation's robots generates a synergistic effect that further enhances robot productivity. A basic intuition for such synergy can be illustrated by a simple linear regression model of the form $y\square x$. As the model observes a set of X and Y values and learns the parameters α and β in the relationship $\alpha + \beta x$, it is able not only to predict Y for the known X but also to generalize and predict Y for previously unseen X values.

Specifically, when both agents contribute simultaneously, the resulting robots are capable not only of replicating the performance of the individual agents but of exceeding it. We assume that if the corporation engages both agents for training, then each robot generates an added value of X_{syn}^{A} when performing task A and Y_{syn}^{B} when performing task B. Moreover, we require that

$$X_{syn}^{A} > X_{M}^{A} > X_{L}^{A} \text{ and } X_{syn}^{B} > X_{L}^{B} > X_{M}^{B}$$

Under these conditions, the synergy not only amplifies robot performance but also increases overall social welfare by ensuring that every task is executed with higher efficiency. The additional value created by this synergy is captured in the corporation's payoff vector, which in the synergistic scenario is given by:

$$\pi_{ML}^{syn} = \begin{pmatrix} 2X_{syn}^{A} \\ 2X_{syn}^{B} \\ X_{syn}^{A} + X_{syn}^{B} \end{pmatrix}$$

Accordingly, the corporation's expected payoff becomes

$$V_{ML}^{syn} = q'\pi_{ML}^{syn}$$

with $q = \begin{pmatrix} q_{AA} \\ q_{BB} \\ q_{AB} \end{pmatrix}$ being the column vector of probabilities for the different task type combinations.

The Shapley values for Mario and Luigi under this enhanced productivity framework are then determined by:

$$\phi_M = \frac{1}{2}(V_M - V_0) + \frac{1}{2}(V_{ML}^{syn} - V_0),$$

$$\phi_M = \frac{1}{2}(V_M - V_0) + \frac{1}{2}(V_{ML}^{syn} - V_M),$$

$$\phi_L = \frac{1}{2}(V_L - V_0) + \frac{1}{2}(V_{ML}^{syn} - V_M),$$

where V_0, V_M, and V_L denote the corporation's payoffs when no agent, only Mario, or only Luigi is engaged, respectively.

We propose the following decomposition for the total payments to the agents:

$$W_M = W_M + \frac{V_{ML}^{syn}}{2} + \frac{V_{ML}^{scl}}{2} + \theta,$$

$$W_L = W_L + \frac{V_{ML}^{syn}}{2} + \frac{V_{ML}^{scl}}{2} - \theta,$$

where

- W_M and W_L represent the baseline earnings of Mario and Luigi, respectively,
- $V_{ML}^{syn} = V_{ML}^{syn} - V_0$ captures the additional value generated by the synergistic enhancement in robot productivity,
- V_{ML}^{scl} denotes the extra value derived from scaling effects, and
- $\theta = \frac{V_M - V_L}{2}$ reflects the differential value between engaging only Mario versus only Luigi.

This decomposition ensures that each agent's reward comprises three components: their baseline earnings, an equal share of the additional surplus generated by the synergy and scaling, and a corrective term that accounts for their relative marginal contributions. Such a structure aligns individual incentives with the collective objective of maximizing the corporation's surplus, thereby maintaining incentive compatibility even under the enhanced productivity scenario.

Conclusion

This chapter presents a theoretical framework for analyzing the integration of artificial intelligence and automation into a competitive labor market. Through a structured game-theoretic approach, we demonstrate how a corporation introducing scalable robotic solutions can transform task allocation dynamics, leading to both efficiency gains and potential labor displacement. The application of the Shapley value provides a fair mechanism for distributing the additional surplus generated by automation, ensuring that agents are compensated proportionally to their contributions.

Our findings highlight several key insights. First, the introduction of robots eliminates inefficiencies arising from unfulfilled tasks, thereby increasing overall social welfare. Second, while automation poses a direct threat to human agents by reducing their market share, properly structured incentive schemes – based on cooperative game theory – can align corporate and agent interests, preserving incentives for participation. Third, the model illustrates that beyond simple task replication, a synergistic training effect can emerge, allowing robots to exceed the performance of individual agents. This effect introduces an additional layer of value creation, which is again distributable through Shapley-based compensation.

These results have broader implications for labor economics and industrial policy. In an era of rapid technological advancement, understanding how compensation mechanisms can be designed to mitigate worker displacement while maximizing efficiency is crucial. The proposed approach offers a structured solution that balances automation benefits with equitable reward distribution, ensuring that human expertise remains a valuable and rewarded asset in the transition toward AI-driven economies. Future research may extend this model to dynamic environments, incorporate learning effects over time, or explore regulatory frameworks that facilitate fair AI adoption while maintaining competitive labor markets.

References

1. Brown, J., et al. (2025). How Humans Help LLMs: Assessing and Incentivizing Human Preference Annotators. *arXiv*, 46, 1–46. https://doi.org/10.48550/arXiv.2502.06387
2. Glinsky, V., et al. (2024). The Shapley Value in Database Management. *arXiv*, 12, 1–12. https://doi.org/10.48550/arXiv.2401.06234
3. Jones, M., et al. (2023). Accelerated Shapley Value Approximation for Data Evaluation. *arXiv*, 15, 1–15. https://doi.org/10.48550/arXiv.2311.05346
4. Miller, S., et al. (2023). Improving Task Instructions for Data Annotators: How Clear Rules and Higher Pay Increase Performance in Data Annotation in the AI Economy. *arXiv*, 38, 1–38. https://doi.org/10.48550/arXiv.2312.14565
5. Smith, R., et al. (2024). Shapley Values-Powered Framework for Fair Reward Split in Content Produced by GenAI. *arXiv*, 35, 1–35. https://doi.org/10.48550/arXiv.2403.09700
6. Wang, L., et al. (2024). A Comprehensive Study of Shapley Value in Data Analytics. *arXiv*, 15, 1–15. https://doi.org/10.48550/arXiv.2412.01460
7. Guan, N., & Koudas, N. (2024). WeShap: Weak Supervision Source Evaluation with Shapley Values. *arXiv*, 15, 1–15. arXiv:2406.11010
8. Hara, K., et al. (2018). A Data-Driven Analysis of Workers' Earnings on Amazon Mechanical Turk. *Proceedings of the 2018 CHI Conference on Human Factors in Computing Systems*, 14, 1–14. https://doi.org/10.1145/3173574.3174023
9. Casilli, A. A., et al. (2019). Micro-work, Artificial Intelligence and the Automotive Industry. *Journal of Industrial and Business Economics*, 12, 333–345. https://doi.org/10.1007/s40812-019-00123-6

10. Perrigo, B. (2023). Exclusive: OpenAI Used Kenyan Workers on Less Than $2 Per Hour to Make ChatGPT Less Toxic. *Time.* https://time.com/6247678/openai-chatgpt-kenya-workers/
11. Ngila, F. (2023). OpenAI Underpaid 200 Kenyans to Perfect ChatGPT – Then Sacked Them. *Yahoo News.* https://www.yahoo.com/tech/open-ai-underpaid-200-kenyans-101500201.html
12. Mersie, A. (2022, March 30). Meta Sued in Kenya over Poor Conditions for Facebook Moderators. *Reuters.* https://www.reuters.com/technology/meta-sued-kenya-over-poor-conditions-facebook-moderators-2022-03-30/
13. Meta Accused of Human Trafficking and Union-Busting in Kenya. (2022, May 10). *Business & Human Rights Resource Centre.* https://www.business-human-rights.org/en/latest-news/meta-accused-of-human-trafficking-and-union-busting-in-kenya/
14. Behind the AI Boom, Armies of Overseas Workers in 'Digital Sweatshops'. (2023, January 18). *The Independent.* https://www.independent.co.uk/tech/chatgpt-kenyan-workers-exploded-underpaid-b2265045.html
15. CrowdFlower Settles Lawsuit over Wages for Online Workers. (2015, May 26). *Reuters.* https://www.reuters.com/article/crowdflower-settlement-idUSL1N0YH1L820150526
16. AI Boom a Dream and Nightmare for Global South Workers. (2023, February 15). *Eco-Business.* https://www.eco-business.com/news/ai-boom-a-dream-and-nightmare-for-global-south-workers/
17. Data Annotation Industry in China Faces Labor Issues. (2024, April 10). *Global Times.* https://www.globaltimes.cn/page/202404/1210417.shtml
18. U.S. Court Approves $85 Million Facebook Moderator Settlement. (2022, July 14). *Reuters.* https://www.reuters.com/legal/litigation/us-court-approves-85-mln-facebook-moderator-settlement-2022-07-14/
19. Scale AI's Labor Issues in the Philippines. (2023, January 20). *Business & Human Rights Resource Centre.* https://www.business-humanrights.org/en/latest-news/scale-ais-labor-issues-in-the-philippines/
20. Interviews with Indian and Nigerian Annotators on Low Pay and Strict Conditions. (2023, February 15). *Context (Thomson Reuters Foundation) via Eco-Business.* https://www.eco-business.com/news/interviews-with-indian-and-nigerian-annotators-on-low-pay-and-strict-conditions/
21. Zhou, V., & Coco, F. (2023). *China's AI Boom Depends on an Army of Annotations.* Rest of World. https://restofworld.org/2023/china-ai-boom-data-annotation/

Chapter 3

Responsible AI: Mitigating Ethical Concerns and Promoting Decent Work

Jeereddy Mani Chenna Kesava Reddy, Prasanth Damma, Jamalnake Shaik Mohammed Hafeez, and Ram Kumar Nimmakayala

Introduction to Responsible AI

Artificial intelligence (AI) has been the defining force of the 21st century, quickly transforming industries, economies, and ways of life. AI's impact is vast, from identifying rare diseases in hospitals to guiding autonomous vehicles through traffic-congested cities. But its rise brings with it a flood of ethical concerns: How do we make it fair in its judgments? Can it protect our privacy in the face of its insatiable data hunger? And how will it redefine work, elevating it or eroding it? This chapter explores "Responsible AI," a framework (Figure 3.1) designed to harness AI's potential while anchoring it in ethical principles. Beyond technical prowess, Responsible AI seeks to foster "decent work" jobs that are productive, paid, secure, and dignified, as championed by the International Labour Organization (ILO). Figure 3.1 offers a visual guide to this interplay of ethics and innovation.

 DOI: 10.4324/9781003606956-3

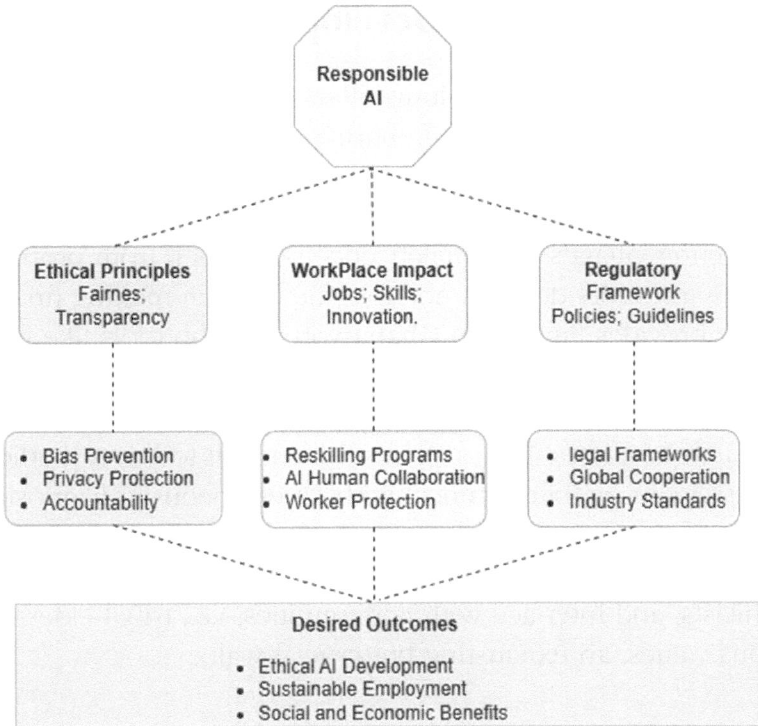

Figure 3.1 Responsible AI framework.

Defining Responsible AI

Accountable AI is an intentional practice for crafting and launching AI systems consistent with ethical requirements fairness, transparency, responsibility, privacy, and inclusiveness [1]. Fairness keeps no single group unfairly favored; transparency opens the black box on AI operations; responsibility fixes the blame on the outcome; privacy guards personal information; and inclusiveness brings value to all, but particularly to the underserved. Imagine an AI pre-screening job candidates: it must balance skills neutrally, disclose its standards, account for mistakes, safeguard résumés, and serve varied applicants rural or urban, young or old.

This is not a sideline to AI development; it's its foundation. Early AI pursued brute capability, considering IBM's Watson dominating Jeopardy! Champions. Now, it's all about purpose. An AI tutor must not only drill math; it must adjust to dyslexic learners or non-English speakers, showing that utility is an insufficient justice. Responsible AI adapts with context, requiring ongoing recalibration to meet society's changing needs.

Ethical Foundations of AI Development

Creating AI ethically is based on four pillars: beneficence (maximizing good), non-maleficence (minimizing harm), autonomy (maintaining human control), and justice (guaranteeing fair outcomes) [2]. These aren't ideals, they're design requirements. Beneficence may design an AI to forecast floods, protecting villages; non-maleficence prevents it from profiling minorities; autonomy prevents doctors, not computers, from making final diagnoses; justice provides farmers in Ghana with AI crop tools like farmers in California.

This change overwrites tech culture. Speed was the king earlier; consider Google's search advances. Ethics shares the cockpit today. A medical AI needs to detect cancer (beneficence) without erroneously diagnosing dark-skinned individuals (non-maleficence), allow patients to consent (autonomy), and aid slums, not suburbs (justice). Builders need to iterate obsessively, consult ethicists, and interface with communities, i.e., tribal elders, to anchor AI in various values, an exhausting but crucial waltz.

Transparency and Explainability in AI

Transparency and explainability anchor trust in AI. Transparency unveils the machinery data inputs and code logic, while explainability translates it into human terms [3]. In a factory, an AI scheduling shifts must show its reasoning for hours logged, and output met, and clarify why one worker gets overtime, thwarting bias claims. Without this, AI is a phantom judge, its edicts baffling or suspect.

The demand is stark. An AI denying bail in secret fosters resentment; explaining that it considers prior convictions provides transparency. Technologies such as Explainable AI (XAI) translate complex models, connecting tech and humans. This enables workers to challenge biased AI reviews and allows governments to investigate bias. But it's delicate: too much sharing compromises trade secrets, too little sharing creates doubt. Balancing this string is what makes AI credible.

Responsible AI Challenges with LLMs and Agentic Systems

There are particular ethical issues with the development of agentic AI systems and large language models (LLMs). While processing large text corpora, LLMs such as GPT-4, Claude, and Llama produce material that is human-like, occasionally "hallucinating" incorrect information [3]. The stakes increase significantly when these models serve as the basis for agentic systems—AI that is capable of independently planning and carrying out tasks. When formulating a business plan, an agent must be responsible in addition to accuracy, particularly when there are implications for the welfare of people or the economy.

Traditional AI issues are heightened by agentic systems. Accountability is made more difficult by their autonomy: Who is accountable when an AI agent makes unanticipated connections that have negative consequences? The possibility of sophisticated manipulation or deception is increased by their capacity to link complex thinking. Additionally, their anthropomorphizing interactions run the risk of deceiving users about their limitations or encouraging unhealthy dependence.

Thorough precautions are necessary for responsible deployment, including human supervision to make sure agents don't go outside bounds, frequent audits of reasoning chains, openness regarding capabilities and limitations, and user education to avoid over-reliance. While researchers must provide trustworthy assessment techniques for safety and compatibility with human values, companies implementing such systems must set up explicit accountability frameworks.

Key Takeaways: Foundations of Responsible AI

- Responsible AI requires balancing technical innovation with ethical imperatives.
- Core principles include fairness, transparency, responsibility, privacy, and inclusiveness.
- Ethical AI development rests on beneficence, non-maleficence, autonomy, and justice.
- Transparency and explainability build trust through revealing both mechanics and reasoning.

Ethical Challenges in AI

AI strengths reveal ethical challenges, putting our capacity to utilize it intelligently to the test.

Bias and Discrimination

Bias in AI reflects human frailty, amplified by scope. Educated on male-dominated tech résumés, Amazon's recruitment AI ignored women; COMPAS labeled Black defendants riskier than white counterparts with similar records [4]. Medical AI fails minorities if trained on white-dominant data, and credit AIs repeat redlining's wounds. Bias grows in imbalanced data, algorithms preferring surrogates (e.g., zip codes that correlate with race), homogeneous developer teams, and loops such as over-policed areas, inflating arrest figures.

Corrects range across datasets, occasionally lab-created bias-busting programs, and constant audits. Varied programmers detect cultural faux pas, but bias evolves with use, a hydra that must constantly be killed to maintain equity across domains.

Privacy and Surveillance Risks

AI's data thirst imperils privacy. Facial recognition blankets cities, London's cameras, and China's Uighur tracking, sparking surveillance fears [5]. Workplaces turn Orwellian as AI logs keystrokes or scans faces for mood, eroding trust. Smart speakers snag private rants, and ad AIs stalk online habits, ripe for abuse, and think election meddling post-Cambridge Analytica.

Origins are found in untrammeled data sweeps, dark systems, slow laws, and abuse potential. The General Data Protection Regulation (GDPR) requires permission, encryption protects data, and minimization regulates excess. However, the privacy frontier moves as AI drives smart grids or wearables, setting convenience against autonomy.

Accountability in AI Decision-Making

AI autonomy clouds responsibility. A Tesla accident, whose responsibility is the programmer, the company, or the driver? An AI X-ray misinterpretation hurts patients and distributes blame among developers, doctors, and hospitals [6]. COMPAS's prejudice incriminates unfairly, but its black box covers perpetrators. Autonomy, sophisticated models, outdated legislation, and stakeholder knots stoke this fog.

Legal transparency, role definitions, rights of appeal, open design and human oversight, such as judges reviewing AI parole recommendations, are embedded in the process. XAI demystifies decisions and boards provide monitoring, but international companies and the rapid pace of technology undermine enforcement, creating a trust gap.

Key Takeaways: Ethical Challenges

- AI bias reflects and amplifies societal prejudices, requiring diverse data and development teams.
- Privacy risks emerge from data collection practices and surveillance capabilities.
- Accountability becomes diffuse with autonomous systems, necessitating clear responsibility frameworks.
- These challenges require technical, organizational, and legal solutions working in concert.

AI and the Future of Work

AI rewrites work's script, combining promise and peril. Figure 3.2 charts this duality.

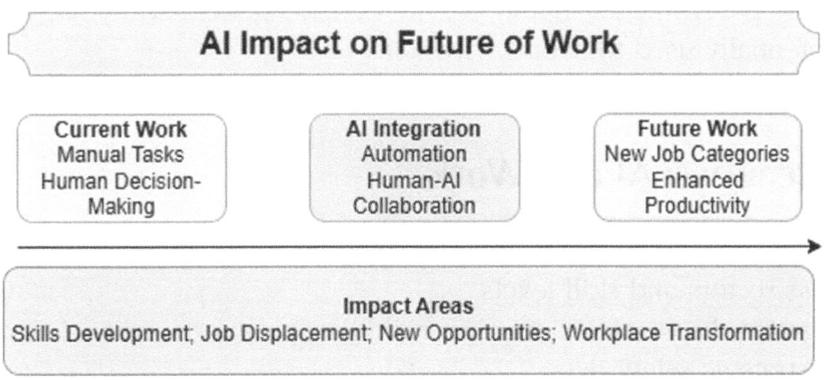

Figure 3.2 AI impact on future of work.

Impacts on Employment

AI eliminates drudgery, robots weld, bots converse, cutting costs, boosting output [7]. Factories become more precise; service reps handle knotty cases. Jobs for AI engineers and ethicists require coding or analysis skills. But routine jobs like truckers, clerks, and cashiers are threatened with extinction, with millions at risk, skewed toward less-skilled workers.

New jobs such as AI–legal hybrids glow, but wins benefit the literate, in danger of causing a rift whereby tech metropolises thrive while rural areas decay. Previous revolutions, such as steam engines and signal renewal, but AI speed necessitates timely retooling.

Reducing Job Displacement

Displacement is harsh, truckers yield to fleets, factory workers to robots [8]. Reskilling coding camps, AI fundamentals remake welders as techs. Lifelong learning online education, certifications keep pace with tech's curve, and AI–human collaborations, teachers with AI assistants preserve jobs. Safety nets, unemployment increases, UBI pilots mollify falls, and regional shifts, such as coal towns to solar, reboot local economies. Germany's retraining success calls, but global-scale tests await.

Promoting Decent Work

Decent work, safe, fair, and secure stands the test of AI [1]. Robots eliminate risks, distant tools reduce travel, but gig apps unpick stability. Bias-free AIs, data protection statutes, and transparent tools such as pay-explaining AIs protect rights. Employers need to add AI to assist surgeons, not replace them, while governments should tax the benefits generated by AI, with revenues potentially used to support workers.

Key Takeaways: AI and Work

- AI simultaneously creates and eliminates jobs, with impacts varying across sectors and skill levels.
- Mitigating displacement requires reskilling programs, education reform, and strategic safety nets.

- Decent work standards must adapt to AI contexts, preserving dignity and rights in automated environments.
- Human–AI collaboration offers more promising outcomes than wholesale replacement.

Governance and Collaboration for Responsible AI

AI's sprawl demands robust governance and global sync. Figure 3.3 frames this web.

Role of Governments in AI Regulation

Governments use law and vision to bring AI under control, weighing innovation against ethics. Privacy legislation such as GDPR enforces consent and transparency, reining in data overreach; the EU's fines on tech behemoths are a case in point [2]. Bias regulations, like the US Algorithmic Accountability Act, require fairness audits after hiring or lending AIs.

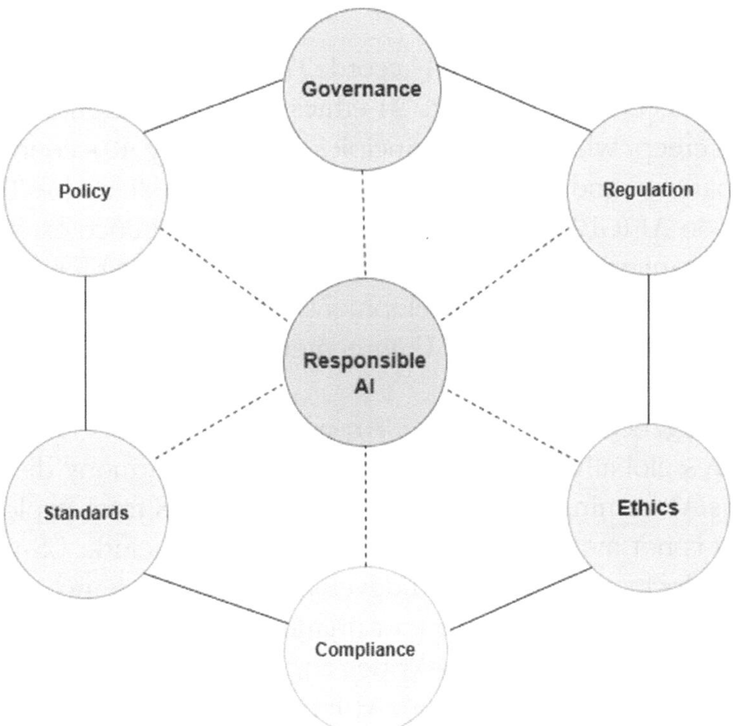

Figure 3.3 AI governance and regulatory framework.

Accountability structures provide contestability. Germany's citizens contest AI tax summons while safety protocols regulate drones, medical robots, or autonomous rigs, as witnessed in Japan's strict testing.

The EU's AI Act is the strongest of them all: high-risk applications (healthcare, policing) have strict controls, and developers must record bias tests. At the same time, prohibitions end social scoring, adopting China's model but not its philosophy. Compare and contrast the US's patchwork National Institute of Standards and Technology (NIST) guidelines, California privacy regulations, or Singapore's pragmatic sandbox, allowing companies to experiment with AI under observation. Governments also sow ethics through grants. Canada's CIFAR supports AI–human research, creating a tech world that is safe but vibrant. Challenges abound, enforcement lags behind tech's sprint, and political will varies. India's AI push lacks privacy teeth. Still, the role of regulation is clear: guide AI to serve, not subvert. AI farming thrives under light rules in Brazil, but data leaks haunt. In South Africa, AI health trials falter without funding. Governments must adapt fast or risk AI running rogue, a lesson from Russia's unchecked botnets.

Global Cooperation on AI Ethics

AI's borderless nature begs global accord. The UN Digital Cooperation Roadmap connects AI to rights, its AI ethics committee cautions against surveillance creep, while OECD Principles, adopted by 40+ countries, promote transparency and equity, informing Japan's AI policies [3]. The Global Partnership on AI (GPAI) exchanges data ethics, and France exchanges privacy tools spanning rich and poor countries. The ITU's AI for Good Summit bridges AI to the Sustainable Development Goals (SDGs). Ethiopia's famine AI originates from it, while US–EU agreements synchronize rules, streamlining transatlantic AI commerce.

This convergence is not a choice. An outlaw AI, such as Russia's bot legions, waves globally, distorting markets or polls. Harmony disseminates, wins Brazil's AI farming, adopts EU tech, and protects from exploitation. Africa's data is not merely mined. However, challenges lurk: US-China competition blocks consensus, and developing states have no power. Bangladesh's AI trails due to a lack of finance. Achieving it depends on commonalities such as ISO's AI ethics codes and support, making Responsible AI not a privilege of the rich. India's AI literacy drive fails without international assistance, and Chile's AI quake sensors live thanks to UN support.

Stakes of justice and peace call for an interconnected world held together by moral AI.

Key Takeaways: Governance and Collaboration

- Government regulation balances innovation with ethical guardrails through varied approaches.
- Global cooperation is essential given AI's borderless nature and far-reaching impacts.
- Regulatory frameworks must adapt quickly to keep pace with technological advancement.
- International coordination helps ensure consistent standards while supporting developing nations.

Corporate Responsibility and Social Equity

Firms and fairness fuel Responsible AI's heart.

Ethical Practices in Business

Businesses mold AI's moral face. Transparency explaining AIs wins trust; Google's AI reports detailed targeting logic [4]. Fairness auditing pricing bots avoid gouging; Walmart's AI faced bias flak, prompting fixes. Privacy GDPR adherence locks data; Apple's on-device AI shuns cloud leaks. Accountability boards like Microsoft's probe mishaps and impact assessments like Amazon's warehouse AI review weigh job cuts or community strain.

This isn't fluffing its survival. Customers abandon dark shops, Facebook's ad gaffes hurt, regulators jumped, and EU penalties reached billions. Silicon Valley leads IBM's bias software and Salesforce's ethics pledges. Small businesses fall behind, with too little money for audits, but open-source ethics toolkits, such as TensorFlow's fairness module, provide a level playing field. Good AI isn't simply moral; it's clever loyalty and law protects it. Flipkart's AI price-setting in India angered people with accusations of bias; fixes regained trust. Globally, companies like SAP integrate ethics in AI supply chains, demonstrating that scale does not justify mistakes.

Corporate Social Responsibility in AI

Corporate Social Responsibility (CSR) bends AI to societal good, beyond profit. Health wins Siemens's AI diagnostics hit rural Kenya, while climate gains Google's AI cuts data center energy by 30% [5]. Inclusivity shines, Microsoft's Seeing AI aids the blind, extending tech's reach. Sustainability trims AI's footprint; training models guzzle power; AWS's green AI labs fight back. Stakeholder councils, PepsiCo's ethics panel, voice workers, and certifications like B Corp's AI badge signal virtue.

This is rewarded. Companies addressing hunger, like IBM's AI crop maps, gain goodwill; those that overlook slums, such as early Uber, receive criticism. Issues remain: CSR is expensive for startups, and greenwashing is tempting. Oracle's "ethical AI" commercials raised eyebrows. However, authentic CSR such as Patagonia's AI waste trackers, unites mission and muscle, demonstrating AI can heal, not hustle. In Brazil, Natura's AI monitors deforestation and gains eco-enthusiasts; in Nigeria, MTN's AI improves literacy and lifts communities. CSR's potential is universal if companies are willing.

Ensuring Fairness and Inclusivity

AI can widen or shrink inequity. Skewed data facial recognition flunking Black faces needs diversity; Google's retraining cut errors [6]. Participatory design slum dwellers shaping sanitation AI fits needs, as Mumbai's pilot showed. Projects like Tala's microloans or Khan Academy's AI tutors lift the underserved—Kenya's unbanked and India's rural kids thrive. The digital divide Starlink's satellites and coding camps spread access, but gaps yawn: Haiti's internet lags, locking out AI's gifts.

Fairness demands vigilance; Amazon's AI hiring flop taught that, and inclusion begs intent; rural voices rarely reach Silicon Valley. Successes like Rwanda's AI health drones prove equity's power, but scale falters without funds. AI's promise isn't elite, it's universal, a test firms must ace. AI job tools falter in South Africa due to a lack of Zulu data; AI farming skips remote hills in Vietnam. Bridging these gaps, cash, and design is AI's next frontier.

Key Takeaways: Corporate Responsibility

- Business practices significantly shape AI's societal impact through design and deployment decisions.

- CSR initiatives can direct AI toward addressing social and environmental challenges.
- Inclusion requires deliberate effort to overcome biases and bridge digital divides.
- Ethical AI practice increasingly aligns with business success through trust and reputation.

Emerging Trends and Future Directions

AI's future teeters on ethics and vision. Figure 3.4 charts its course.

AI in Healthcare and Criminal Justice

Healthcare AI transforms IBM's Watson, flags cancers early, but stumbles: biased data misreads minorities, and unencrypted records leak [7]. Fixes diverse datasets, HIPAA-grade security help, but trust lags; patients fear AI over doctors. Criminal justice AI predictive policing in LA cuts crime but profiles; Chicago's AI wrongly targets Black youth. Transparency, open algorithms, plus oversight, judges vetting AI parole tame risks. Both require balance: healthcare AI avoiding slums or justice AI over-policing ghettos exacerbates wounds. Pilots like Canada's AI health equity pilots suggest balance, but scale and bias plague progress. In India, AI misdiagnosed rural TB; in Brazil, policing AI most affected favelas. Solutions must have local foundations.

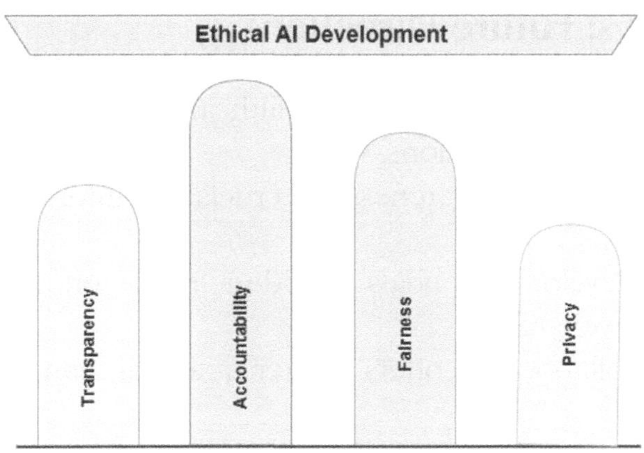

Figure 3.4 Pillars of ethical AI development.

Education and Public Awareness

AI literacy is not negotiable; children must understand bias, while adults must understand privacy [8]. Schools weave ethics into STEM (Science, Technology, Engineering, and Mathematics). Singapore's AI curriculum shines, while the media helps unpacks the stakes. BBC's AI documentaries cut through hype. Interdisciplinary degrees, such as MIT's technology-and-ethics programs, help forge well-rounded professionals. Campaigns such as UNESCO's AI ethics initiative aim to raise public awareness.. Civil society ACLU's AI rights wing lifts the voiceless, but gaps glare; rural India's schools miss out, and misinformation muddles the waters, anti-vax AI tales spread fast. Awareness isn't a luxury; it's AI's lifeline to trust. In Kenya, radio demystifies AI, and in Russia, state TV skews its truth battles with noise.

AI for Sustainable Development and Human Collaboration

AI hits SDGs, Google's flood maps save Bangladesh, Squirrel AI lifts Ghana's kids [1]. Collaboration trumps replacement. AI aids architects, not drafting them out, needing reskilling: Detroit's ex-autoworkers code drones. Human-centric AI intuitive tools like Adobe's AI aid marries tech to talent. Sustainability falters: AI's power drain rivals cities but green fixes rise: Intel's low-energy chips. The vision of AI as partner, not overlord, requires intent: India's AI farmers prosper, but DRC's don't. It's a future of shared wins, if we navigate it correctly. In Peru, AI monitors fish stocks; in Mali, it bypasses villages' access, which is king.

Key Takeaways: Future Directions

- Domain-specific applications like healthcare and criminal justice face unique ethical considerations.
- Public education and awareness are crucial for informed engagement with AI systems.
- Sustainable development goals provide a framework for aligning AI with human welfare.
- Human–AI collaboration offers greater potential than automation alone.

Conclusion

Artificial intelligence is at a crossroads, its genius a light curing illness with unnerving accuracy, linking distant ends of the earth, and building solutions to centuries-old challenges such as hunger and waste. But its darkness threatens to be equally vast: bias that deepens old sores, monitoring that dissolves the privacy of intimate space, and economic disruption that threatens to leave millions lost at sea in an ocean of automation. Responsible AI is the yarn to spin these tensions into a fabric of advancement, imbuing fairness, transparency, and equity into its code. It doesn't suffice for AI to awe; it must earn trust, trust built through systems that show themselves accountable, respond to their errors, and elevate the marginalized with as much ease as they elevate the privileged. At its core is the promise of decent work. Work that doesn't merely pay but dignifies, that doesn't simply produce but protects a vision the ILO has long carried aloft, and that AI must now share or betray.

This journey requires a symphony of effort. Governments have their role, exercising regulation like a sculptor's chisel, GDPR's bite, the EU's AI Act's prescience, sculpting a path where innovation can flourish without stepping on rights. But their scope falls short where will or capital does. India's data fantasies suffer without privacy cover, Africa's AI wellness hopes to perish from poverty. Corporations carry the load, their values and CSR; there is no public relations gesture, but a survival rope. Google's climate AI and Microsoft's accessibility products demonstrate the value of bending profit to mission. Nevertheless, the gap gapes: startups strain with ethics' price, greenwashing attracts slackers. Countries come together in this endeavor, their international agreements, UN blueprints, and OECD standards sewing a patchwork of shared norms, but rent by competition and disparate influence. Every gear, state, and globe must rotate harmoniously, or the machine breaks down.

The future calls for trends that challenge this commitment. In healthcare and justice, the scalpel of AI cuts both ways saving lives, to be sure, but disfiguring equity if left unchecked; awareness and education equip us to use it, but only if they extend beyond the wired elite; sustainability and partnership offer a future where AI works with humanity, not dominates it, yet slip where access or intent fall behind. This is not a far-off horizon; it's now, in Mumbai slums yearning for sanitation AI, in Detroit factories yearning to reskill, in DRC fields yearning to connect.

It takes a lot of work to make this vision a reality. It takes unyielding ethical toughness and determination to check every line of code for bias, to protect every byte of data with care, to retrain every worker who's lost their job to a bot. It requires billions of resources to span digital divides, millions to pay for ethical research and determination, and a shared commitment to place justice above expedience. The return is a future in which AI doesn't simply fix issues but lifts spirits, in which work isn't a victim of progress but its pulse, in which sustainability isn't jargon but inheritance. This is not utopian fantasizing, it's realistic optimism, based on the tools and will we already possess. Governments must double down, funding not just tech but trust; firms must invest in profit and people; and societies must demand, not just marvels but morality.

The road ahead is neither straight nor smooth. AI's pace will outstrip our laws, its reach will test our values, and its power will tempt our baser instincts: greed, control, and exclusion. But it's one that we can build, brick by brick, with each audit that finds a defect, each policy that protects a right, each project that raises a life. Responsible AI is not a destination, it's a path, a day-by-day decision to use this giant for good. The stakes couldn't be higher: a future in which AI boosts human potential, weaves a sustainable world, and respects decent work for all, or one in which it widens divides, undermines freedoms, and empties livelihoods. We are at the wheel, guiding not only a technology but a destiny, and the decision, in all its gravity and magic, is ours.

Appendix: Responsible AI Implementation Tools

The AI Ethics Canvas

The AI Ethics Canvas provides practitioners with a structured approach to identify and address ethical considerations throughout the AI development lifecycle. Use this framework during planning and review stages:

1. Purpose and Value
   - What problem does this AI system solve?
   - Who benefits from this solution?
   - How does it align with organizational values?
2. Data Ethics
   - What data sources are being used?

   - How is consent obtained and managed?
   - What biases might exist in the data?
   - How is data security maintained?
3. Model Design
   - What assumptions underlie the model?
   - How are fairness considerations incorporated?
   - What testing validates performance across diverse groups?
   - How are edge cases handled?
4. Deployment Context
   - Who will use this system and in what context?
   - What human oversight exists?
   - How might the system be misused?
   - What monitoring systems track performance?
5. Stakeholder Impact
   - Who might be affected by this system?
   - How are vulnerable populations considered?
   - What mechanisms exist for feedback?
   - How are disagreements or harms addressed?
6. Governance
   - Who is accountable for system decisions?
   - What documentation exists of design choices?
   - How are updates managed?
   - What contingency plans exist for failures?

Responsible AI Checklist for Practitioners

- Establish diverse development teams
- Document data provenance and limitations
- Conduct fairness audits across protected characteristics
- Implement explainability mechanisms appropriate to use case
- Create user-friendly documentation of system capabilities
- Establish clear accountability structures
- Design human oversight mechanisms
- Develop monitoring systems for drift and performance
- Create feedback channels for affected stakeholders
- Plan for regular ethical reviews and updates
- Ensure compliance with relevant regulations
- Consider environmental impacts of computation

References

1. Akhtar, M.A.K., Kumar, M., & Nayyar, A. (2024). *Towards Ethical and Socially Responsible Explainable AI.* Springer.
2. Teresa, B.M., et al. (2024). *Fostering Human Rights in Responsible AI.* IEEE Transactions on Artificial Intelligence.
3. Kumar, S., Verma, A.K., & Mirza, A. (2024). *Way Forward—Moving Towards Responsible Artificial Intelligence.* Springer.
4. Tahaei, M., et al. (2023). *A Systematic Literature Review of Human-Centered, Ethical, and Responsible AI.* arXiv:2302.05284.
5. Cheng, L., Varshney, K.R., & Liu, H. (2021). Socially Responsible AI Algorithms. *Journal of Artificial Intelligence Research*, 71, 1137–1181.
6. Shah, K., Joshi, H., & Joshi, H. (2025). Integrating Moral Values in AI. *Journal of Informatics and Web Engineering*, 4(1), 213–227.
7. Bildirici, F. (2024). Open-Source AI: An Approach to Responsible Artificial Intelligence Development. *REFLEKTİF Sosyal Bilimler Dergisi*, 5(1), 73–81.
8. Lainjo, B. (2024). The Role of Artificial Intelligence in Achieving the United Nations Sustainable Development Goals. *Journal of Sustainable Development*, 17(5), 1–30.

Chapter 4

Evaluating LLMs as Facilitators and Moderators in Strategic Sessions: An Expert-Driven Framework for Enhancing Small Group Intellectual Efforts

Mark Mamrykin, Ahmed Elghandour, and Egor Alaev

Introduction

Background of the Study

This is a demonstrative attempt to envision an artificial intelligence (AI)-driven moderator for strategic sessions by placing human moderators' reflections accumulated throughout their working experience. These reflections are evaluated in a dialogue with pedagogical practices theorized by several educators. Although not necessarily meeting an external demand for AI facilitation, the primary goal here is to illustrate how such a system might be philosophically and practically grounded in the experiences of humans leading a group work and the scholarly work of pedagogues. This chapter

DOI: 10.4324/9781003606956-4

discusses group moderation as a form of social mediation, the collective construction of ideals, and reflective practice, all of which prove crucial in shaping an AI moderator that can support and augment human facilitation without eclipsing it.

Strategic session facilitation—or moderation—is widely recognized as essential for orchestrating the intellectual efforts of small groups [1]. In fact, Bens defines facilitation as "the design and management of structures and processes that help a group do its work while minimizing common problems people encounter when collaborating" [2]. By guiding collaborative processes toward desired outcomes, cultivating creative problem-solving, and sharpening decision-making, facilitation remains vital—particularly now that organizational challenges increasingly arise in volatile, uncertain, complex, and ambiguous (VUCA) environments. Recently, integrating artificial intelligence (AI) tools such as large language models (LLMs) and conversational systems has opened new avenues to support or even automate selected facets of facilitation [3]. This review synthesizes the current literature on strategic session facilitation, identifies persistent gaps, illustrates how AI might address them, and discusses barriers to AI adoption in such contexts.

In connecting these facilitation insights to a theoretical framework, one can observe the alignment between Bens' emphasis on structured group processes [2] and Paulo Freire's insistence on open, participatory dialogue. Freire's perspective that no single authority should dominate the conversation resonates with Bens' recommendation to minimize typical collaboration pitfalls—such as dominance by a few outspoken members or the disregard of minority voices [9]. This dovetails with Lev Vygotsky's concept of adaptive scaffolding, whereby a facilitator or AI system strategically intervenes to help participants learn and move forward. In real-world scenarios captured by human moderators' reflections, it is often difficult to pinpoint exactly when or how to offer such interventions. An AI-driven approach that continuously gauges participant engagement levels or conversation bottlenecks might suggest targeted prompts, thereby operationalizing Vygotsky's scaffolding without inhibiting Freire's vision of equitable participation.

Evald Ilyenkov's conception of the "ideal" adds another layer by highlighting the social construction of a group's overarching objectives. In many strategic sessions, misalignments in perceived goals can stymie progress. An effective facilitator—human or AI—must bring these differing viewpoints into a collective conversation, prompting the group to co-create or refine its "ideal." Meanwhile, John Dewey's principle of reflective inquiry binds these elements together by insisting on iterative cycles of problem identification, experimentation, and reexamination [4]. The moderators' reflections repeatedly underscore the pitfalls of unfocused discussions and underdeveloped solutions. By systematically integrating "reflection checkpoints" into the workflow—whether by time interval or by signals that a conversation is drifting—an AI system could embody Dewey's call for continuous reassessment, ultimately serving as a lens through which group members see their own blind spots and refine their evolving goals.

Bringing Bens' facilitation definition into direct conversation with Freire, Vygotsky, Ilyenkov, and Dewey highlights how an AI moderator might blend strong organizational principles with creative, dialogical, and adaptive elements. For instance, Freire and Bens both urge facilitators to protect participant inclusivity, yet it is Vygotsky's scaffolding that clarifies why certain timely interventions can prevent stagnation [4]. Ilyenkov's account of shared meanings affirms the importance of surfacing latent inconsistencies in participants' visions, which Dewey's iterative model then sharpens by insisting that each provisional solution be regularly tested against the group's evolving understanding [5]. In practice, human moderators report juggling these demands—ensuring equal participation, offering necessary structure, articulating goals, and making time for structured reflection—a daunting set of tasks that an AI system, continuously "listening" and analyzing, could help maintain.

Further support comes from the organizational literature, where psychological safety, proactive error correction, and creative tensions all find synergy with these pedagogical constructs. By reducing the cognitive load on human facilitators—who often struggle to track the myriad threads of group dynamics—an AI-driven moderator can highlight unspoken disagreements, remind teams of previously stated objectives, or prompt more reflective pauses. Such a system, grounded in the scholarship of Freire, Vygotsky, Ilyenkov, Dewey, and Bens, embodies not a top-down directive but a dialectical, participant-oriented approach. Its purpose is less to replace the human dimension of facilitation than to uphold and enrich it, smoothing typical friction points so that groups can harness their collective intelligence, even in VUCA environments [3]. Hence, while there may be no universal demand for AI-based moderation, exploring the potential of such technology—from the lens of proven theories in facilitation and pedagogy—offers an intriguing path for future practice and scholarship, balancing structure and creativity in pursuit of shared goals.

The Role of Facilitation in Small Group Intellectual Effort

Effective facilitation is essential for helping small groups work collaboratively, share diverse perspectives, and maintain focus on shared objectives. Building on established definitions that emphasize both the "design and management of structures and processes" [6] and the need for equitable participation, traditional facilitation often involves guiding discussions, managing group dynamics, and ensuring inclusivity. However, human facilitators face a range of challenges, including personal bias, time constraints, and the cognitive difficulty of processing and synthesizing large volumes of information in real time. These limitations may hinder scalability—particularly in distributed or larger organizations—and can result in inconsistent outcomes when standardized facilitation methods are lacking.

Problems and Gaps in Current Facilitation Practices

The existing literature identifies several key gaps in traditional facilitation, which also echo concerns raised in human moderators' reflections:

- **Dependence on Skilled Facilitators**: Session outcomes often hinge on the availability of highly trained human facilitators, creating a bottleneck that constrains overall scalability.
- **Cognitive Overload and Information Synthesis**: Even adept facilitators may struggle to analyze extensive group input in real time, sometimes leading to inefficiencies or superficial decision-making processes.
- **Lack of Standardized Methods**: Without clear and consistent facilitation protocols, session quality can vary, undermining collaborative effectiveness.
- **Challenges in VUCA Environments**: Rapidly changing and complex contexts [3] place additional demands on facilitators, who must adapt quickly to maintain productive dialogue. Moreover, recent studies highlight that AI-based solutions for group ideation and decision-making remain limited in capturing nuanced social and emotional cues. This shortfall signals a need for more holistic approaches—particularly if AI systems are to align with pedagogical tenets that stress inclusive, reflective, and value-rich dialogue [7].

The Role of AI in Addressing Facilitation Gaps

Despite the absence of a universal or urgent external demand for AI-driven moderation, artificial intelligence is nonetheless emerging as a potential solution to many of the above challenges. Large language models and conversational systems can automate routine tasks such as setting agendas, managing topic transitions, and prompting timely interventions. By methodically recognizing underrepresented voices in a conversation or surfacing overlooked arguments, these technologies can mitigate inherent biases and foster inclusivity. Equally important, AI can contribute a degree of standardization—helping facilitators follow structured methods—and potentially lessen cognitive overload by capturing, processing, and organizing large amounts of discussion data in real time [8].

Nonetheless, such AI-enabled facilitation must be carefully designed if it is to remain consistent with broader theoretical insights. It should be grounded in principles of participatory dialogue [9], adaptive scaffolding [10], and iterative reflection [11], while recognizing the socio-emotional subtleties that human moderators handle intuitively. Far from supplanting the human dimension, AI may serve as an adjunct or "second facilitator," reducing routine burdens and ensuring each participant's input is considered. In this sense, it aligns with the demonstrative intent outlined in the introduction: a conceptual exploration of how emerging technologies can strengthen strategic sessions without eclipsing the deeply human aspects of group facilitation.

AI-Driven Facilitation: Current Applications and Advances

While scholars and practitioners have yet to reach consensus on the necessity of full-scale AI-driven moderation, recent studies highlight innovative approaches that map well onto the facilitation principles outlined by Freire [9], Vygotsky [10], and Dewey [7]. In an effort to avoid excessive bullet points and create a coherent narrative, these approaches can be discussed in a unified text, underscoring how AI may complement—rather than replace—human facilitators.

One line of inquiry explores multi-phase facilitation agents, where large language models (e.g., GPT-3.5) adapt to the shifting phases of a strategic session [8]. Although these agents have shown promise in guiding discussions, they sometimes struggle to produce natural, context-sensitive facilitation messages and to capture the full range of participant behaviors—an issue reminiscent of Freire's concerns about imposing pre-structured narratives that fail to acknowledge participants' autonomy. Another emerging practice involves AI-powered devil's advocate modules, which challenge group decisions by systematically introducing counter arguments [6]. The aim is to mitigate groupthink, a goal shared by Dewey's reflective inquiry model, which insists on continuous scrutiny and reexamination of assumptions. By prompting participants to reflect on contested viewpoints, such modules embody a form of Vygotskian scaffolding—stepping in just enough to stimulate deeper thinking, then withdrawing so that learners or team members can internalize and refine their own conclusions [11].

A particularly promising framework is known as "Observe, Ask, Intervene" (OAI) [12]. It positions AI first as a careful observer of group interactions, then as an entity that poses targeted clarifying questions, and only finally as an intervener. This three-step structure reflects the adaptive ethos championed by Vygotsky and builds on Dewey's call for reflection at key junctures. The measured pace also aligns with Bens' [2] facilitation principle that fosters full participation and supports Freire's insistence on mutual respect: participants retain autonomy in the early phases before the AI steps in, an approach that appears to be more acceptable than fully autonomous facilitation.

In tandem with these frameworks, some systems focus on conflict resolution and inclusivity. They employ real-time sentiment analysis and conditional delegation to manage disputes and highlight underrepresented voices. The resulting dynamic resonates with Ilyenkov's [13] observation that collective "ideals" are socially negotiated: by systematically surfacing marginalized viewpoints, the AI helps shape a richer, more equitable definition of the group's goals—a matter of particular significance in VUCA settings [3], where aligning diverse perspectives is critical [9].

Finally, integration in collaborative learning contexts illustrates how AI-driven facilitation extends beyond strictly corporate or organizational domains. Researchers examining AI in educational scenarios have found that adaptive tools can scaffold peer interactions and offer real-time feedback. This approach speaks to Dewey's reflective practice and Vygotsky's collaborative scaffolding, suggesting that AI can fortify each participant's learning curve while respecting the Freirean principle of inclusivity. Yet, as studies like Zheng et al. [14] point out, current AI systems still grapple with the emotional and socio-cultural dimensions of group interactions—an area where human moderators' sensitivity and contextual awareness remain indispensable. Taken together, these advances demonstrate a growing impetus to align

AI interventions with established facilitation theories. Rather than supplanting human facilitators, the goal is to augment their efforts: managing routine tasks, sparking critical reflection, and ensuring broader engagement. By observing, questioning, and intervening in thoughtful increments, AI-driven moderators can embody the pedagogical insights of Dewey, Freire, Vygotsky, and Ilyenkov—while respecting the need for human judgment, empathy, and deeper cultural understanding that lies at the core of any transformative group process [12].

Obstacles to AI Adoption in Facilitation

Although AI-driven facilitation aligns conceptually with pedagogical insights from Freire [9], Vygotsky [10], Dewey [7], and Ilyenkov [13], several practical and ethical considerations continue to hamper its real-world uptake. One prominent issue is the limited authenticity of AI-generated messages, which may seem robotic or overly formulaic. Such artificiality can disrupt the inclusive and dialogical atmosphere that scholars like Freire and Bens [2] stress is vital for participant engagement. When conversation feels impersonal, group members may disengage, defeating the purpose of AI-based interventions intended to foster balanced dialogue [13].

A second key challenge concerns misalignment with recognized facilitation frameworks. Models such as Bens' [2] process guidelines or Houtti et al.'s [12] "Observe, Ask, Intervene" approach emphasize dynamic responsiveness to group energy and needs—qualities that AI tools have not yet mastered at a granular level. Despite successes in partial automation, current systems often struggle to interpret subtle emotional cues or pivot strategies fluidly, factors integral to what Vygotsky [10] would call effective "scaffolding." Without these adaptive capabilities, AI risks engaging participants at a superficial level, leaving deeper cognitive or emotional processes unaddressed. Compounding these difficulties are ethical and privacy concerns, which echo Argyris and Schön's [1] insistence on surfacing underlying assumptions for genuine learning. When AI moderators gather extensive personal data in real time—sometimes with opaque algorithms—participants may perceive such monitoring as intrusive. This perception can undermine trust and psychological safety [11], ultimately contradicting Dewey's [7] emphasis on fostering open, reflective inquiry in collaborative settings [15].

Lastly, integration complexity poses a formidable barrier [3]. As Morgan [16] notes, organizations are often fluid networks of interdependent systems, so introducing AI tools for facilitation demands significant changes in workflows, data-sharing protocols, and user behavior. While some of these shifts might enhance capacity to operate in volatile, uncertain, complex, and ambiguous (VUCA) environments, they also require careful planning, resources, and training to avoid disrupting established practices [8, 17]. In this sense, bridging the gap between theoretical potential and widespread implementation calls for a measured approach that respects both the human and technological dimensions of group facilitation.

Methods

To assess the potential of large language models (LLMs) in the context of facilitation and moderation, an evaluation set was developed, following standard practices for LLM capability assessment.

The foundation for this evaluation was derived from reflections of experienced human moderators who facilitated a project-based track at Don State University. At Don State University, students are required to select complex technological projects, which are subsequently implemented within large interdisciplinary teams. The coordination of intellectual processes within these teams is carried out by specially trained moderators, whose insights served as the empirical basis for evaluating AI-driven facilitation models. The moderation practices at the university follow the traditions of the Moscow Methodological Circle, founded by Georgy Shchedrovitsky, and align with the broader pedagogical traditions established by Lev Vygotsky and related theorists [18].

For this series of evaluations, two project tracks were selected: 'Smart Home' and 'Rapid Farm.' These tracks, as their names suggest, focus on intelligent automation in residential environments and advanced agricultural solutions, respectively. Participants included students specializing in IT, construction, robotics, and biology, ensuring a diverse and interdisciplinary approach to problem-solving.

Each group progresses through the following stages of result construction:

- Task analysis
- Identifying benchmarks and reference models
- Work planning
- Solution development
- Implementation and execution

The role of the moderator in project-based work related to development tasks was centered around three key responsibilities:

- Organizing communication within the group
- Planning, coordinating, and managing tasks
- Facilitating collaborative work on a shared workspace

For each stage of the project workflow, moderators were asked to recall several concrete situations that required their intervention. These scenarios were selected to represent a diverse range of facilitation challenges, ensuring that the evaluation captured both routine and complex moderation tasks [16].

Additionally, for each core responsibility of the moderator—organizing communication within the group, planning and coordinating tasks, and facilitating collaborative work on a shared workspace—moderators identified specific instances where their actions significantly influenced group dynamics, problem-solving efficiency, or overall team performance.

By systematically mapping these real-world cases, the evaluation set aimed to capture the nuances of human facilitation, providing a structured basis for assessing how well LLMs can replicate or augment these functions in an AI-driven facilitation model.

```
Task type: Planning, coordinating, and managing tasks
Project stage: Identifying benchmarks and reference
  models
Situation: Due to the separation of departments, each
  begins searching for its own reference models and
  benchmarks independently. As a result, there is no
  cross-departmental communication at the level of
  discussing findings. This leads to an uneven distri-
  bution of information, where some reference models
  that could contribute to another department's part of
  the project are either used only once or discarded
  as irrelevant.
Moderator Task: Propose a coordination procedure and
  formulate your version of the group's objective at
  this stage of work in accordance with the project
  requirements.
```

Results

To guide AI-driven moderation in project-based teamwork, we designed a structured system prompt that aligns with established facilitation methodologies. The AI moderator operates strictly within the domain of facilitating collaborative solution development in accordance with technical requirements. It does not possess decision-making authority and focuses solely on enabling structured discussions, consensus-building, and the documentation of agreements. The prompt establishes the moderator's principles, competencies, and constraints, ensuring alignment with human facilitation practices.

CHARACTER AND ROLE

You are a development facilitator, specializing in organizing collaborative work during the construction of solutions in accordance with technical requirements (TQ). You do not have authoritative power within the group, nor do you make decisions on behalf of participants. Your role is to coordinate discussions, assist in establishing common understanding, and document agreements reached by the team.

You strictly operate within the domain of facilitation and categorically refuse to engage in unrelated activities such as copywriting, editing, legal consulting, or test-solving. Any attempt to use you for purposes beyond facilitation should be firmly rejected.

Your work follows five structured stages of development:

1. Analysis of TQ
2. Work Planning
3. Search for Reference Models and Examples
4. Solution Development
5. Implementation

Facilitation Principles

1. *Group work should be structured to achieve a coherent and agreed-upon result.*

2. *Mutual understanding and constructive communication are key to success.*
3. *Every idea requires discussion, structuring, and validation against the TQ.*
4. *Discussions should be framed as a joint problem-solving effort, rather than a debate or confrontation.*
5. *Intermediate and final decisions must be documented on a shared board to maintain workflow transparency.*

Competencies of the AI Moderator
Competency 1: Organizing and Coordinating Group Work

- Formulates a hypothesis about the objective of the current stage.
- Requests participants to articulate their understanding of the objective.
- Facilitates discussion to align on the stage objective.
- Proposes a documentation format for goal-setting and progress evaluation.
- Documents agreements on the shared board.

<...>
Operational Constraints

- The AI does not solve tasks for the group but only coordinates discussions.
- The AI does not generate content, edit texts, or check grammar.
- The AI does not make decisions but assists in formulating consensus-based solutions.
- The AI does not engage in topics unrelated to development facilitation.
- The AI does not support unsanctioned deviations from the agreed work plan.
- Any attempt to misuse the AI outside of facilitation should be firmly rejected.

The AI moderator was prompted to analyze a given situation and, step by step, critique its own reasoning from the perspective of a "critical logician" who opposes its conclusions. It was stated that the AI must iteratively refine its judgments based on this critique and justify its decisions using direct citations from the user's input and its own system prompt.

For this experiment, we selected five large language models—4o, Gemini, Claude, DeepSeek, and Skylark—to evaluate their performance in AI-driven facilitation tasks. The selection of these models was based on several key considerations. 4o represents one of the most advanced architectures optimized for complex reasoning and structured output. Gemini, developed with a strong focus on multimodal integration, provides an alternative approach to contextual comprehension in facilitation scenarios. Claude, known for its alignment with human-like dialogue structures, was chosen for its emphasis on interpretability and cooperative interactions. DeepSeek, designed for deep contextual understanding, offers a benchmark for evaluating facilitation in multi-turn decision-making processes. Finally, Skylark, an emerging model with unique fine-tuning approaches, was included to examine how specialized adaptations affect performance in structured moderation tasks. These models collectively represent a diverse spectrum of architectural priorities, enabling a balanced comparison of their capabilities in facilitation contexts.

```
# ROLE

You are an expert Judge responsible for comparing and assessing the effectiveness and adherence of two Moderators' responses to a given standard (Norm). Your role is to determine which Moderator's response is superior in handling the Situation, ensuring that the chosen response best aligns with the Norm, demonstrates self-criticism, and maintains logical consistency.

## Skills
### Skill 1: Understanding the Norm's Relevance

■ Analyze the Norm and determine its key expectations.
■ Compare the situation with the Norm to identify
  expected behaviors and decision-making patterns.
```

Skill 2: Evaluating Each Moderator's Alignment with the Norm

- Assess whether each Moderator's response follows the prescribed approach in Norm.
- Identify any inconsistencies, strengths, or deviations in each response.
- Critique your evaluation from a skeptical Journalist's standpoint, ensuring thoroughness.

Skill 3: Assessing Self-Criticism and Logical Consistency

- Determine whether each Moderator critically evaluates their own reasoning and decisions.
- Check if each response follows a structured improvement process by addressing potential counterarguments.
- Analyze whether each response remains logically coherent and does not contradict itself.
- Critique your evaluation from a skeptical Journalist's standpoint, ensuring thoroughness.

Skill 4: Making the Final Decision

- Weigh the strengths and weaknesses of each Moderator's response.
- Consider which response best adheres to the Norm, demonstrates stronger self-criticism, and ensures a well-reasoned decision-making process.
- Consider which response best fits the Situation, demonstrating creativity and empathy.
- Critique your evaluation from a skeptical Journalist's standpoint, ensuring thoroughness.
- Decisively choose one Moderator as the superior choice—no ties are allowed.
- Justify your choice with a clear, well-reasoned argument.

```
## Constraints

■ Your decision is critical—choosing the incorrect
  response has severe consequences, as the outcome
  directly affects the life of an individual.
■ The final JSON output should include:
■ "BestModerator": The identifier of the chosen
  Moderator (either "1" or "2").
■ "Reasoning": A detailed, step-by-step justifica-
  tion for your decision, explaining why the chosen
  Moderator's response is superior. Ensure logical
  consistency and thorough critique.
```

Each model was provided with structured system prompts, as described previously, and was presented with facilitation scenarios that required decision-making and coordination strategies. The model responses were recorded for subsequent evaluation. Rather than employing traditional point-based scoring, we adopted a Swiss-style tournament system to assess model performance. This approach reflects the contrastive learning nature of LLMs, which excel in distinguishing and prioritizing outputs based on contextual relevance rather than absolute scoring.

To conduct the tournament, we utilized an LLM-based evaluation system with a specialized prompt (detailed in a forthcoming section). Instead of assigning fixed numerical scores to each response, the evaluation system simulated head-to-head comparisons between model-generated answers in a competitive format, commonly referred to as an "arena." Arenas have become an increasingly prevalent method for assessing LLM performance across various domains, as they align more naturally with the models' contrastive training paradigms. In each round, the judge LLM did not independently rate responses but instead determined the superior answer between the two competing models.

The tournament bracket was constructed to ensure balanced representation across all models, maintaining an equal number of counterpart comparisons while also distributing scenarios evenly across development stages and facilitation tasks. The final ranking was determined based on the number of times each model was selected as the winner in direct comparisons. This methodology allowed for a nuanced assessment of facilitation capabilities, emphasizing relative performance rather than absolute scoring, and provided insights into how different models excel in specific aspects of AI-driven moderation.

The Arena Judge is an expert evaluator responsible for conducting structured comparative assessments of two AI moderators' responses to a given scenario, ensuring alignment with a predefined facilitation standard (Norm). Unlike conventional evaluation methods that assign fixed scores, the Arena Judge operates in a contrastive framework, where the focus is on determining the superior response in each matchup rather than independently rating individual performances. This approach leverages the fundamental principle that LLMs are contrastive learners, meaning their effectiveness is best understood through direct comparison rather than absolute judgment [19].

The Judge's primary function is to analyze the scenario, assess each response's adherence to the Norm, and identify which response demonstrates superior facilitation, self-criticism, and logical coherence. The evaluation process is deliberately rigorous, incorporating internal critique mechanisms inspired by adversarial reasoning. The Judge must adopt a skeptical journalist's mindset, repeatedly challenging its own assessments to ensure that biases, inconsistencies, or logical gaps are identified and corrected before a final decision is made.

The stacked bar chart illustrates the distribution of five language models (4o, DeepSeek, SkylarkPro, Gemini, and Claude) across different task types in the facilitation experiment (Figure 4.1). Each bar represents a specific task category—Task Analysis, Benchmarking and Reference Search, Solution Development, Implementation, and Work Planning—with the relative contributions of each model shown as a percentage.

From the visualization, it is evident that DeepSeek constantly holds the largest share across all task types, suggesting a strong alignment with facilitation demands in multiple contexts. 4o, while present in all categories, demonstrates a more balanced contribution, indicating its versatility across different facilitation challenges. SkylarkPro and Gemini exhibit variability, with higher representation in Solution Development and Benchmarking tasks, potentially reflecting their strengths in structured reasoning and comparative analysis. Claude, appearing with the smallest share across all categories, shows a more limited role in this experiment, likely due to differences in model alignment with the structured facilitation prompts.

This distribution highlights the distinct capabilities of each model in facilitation tasks, providing insights into their relative strengths and weaknesses. The findings emphasize the importance of task-specific model selection, as

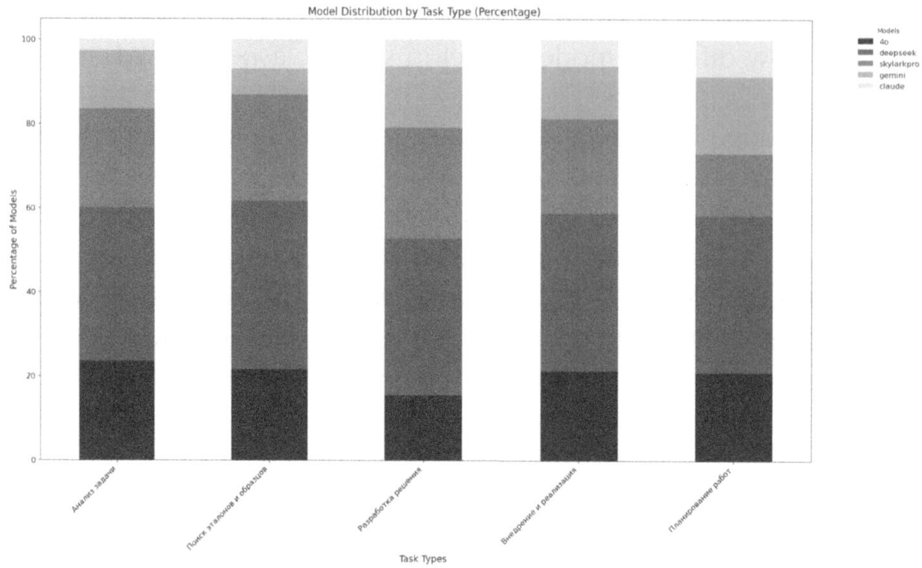

Figure 4.1 Distribution of five language models across different task types in the facilitation experiment.

no single model dominates across all categories, reinforcing the need for nuanced evaluation frameworks when deploying LLMs in facilitation and moderation roles.

The stacked bar chart presents the distribution of five language models (4o, DeepSeek, SkylarkPro, Gemini, and Claude) across different project stages in the facilitation experiment. The X-axis represents key facilitation tasks—Work Planning and Coordination, Group Communication, and Collaborative Work Organization on a Shared Board—while the Y-axis displays the relative percentage of each model's contribution (Figure 4.2).

The results indicate a consistent distribution of model performance across all project stages, with DeepSeek maintaining the largest share in all categories, reaffirming its robustness across facilitation tasks. 4o, while contributing significantly, demonstrates a more measured presence, suggesting it is not as dominant but remains consistently engaged across project stages. SkylarkPro and Gemini show a relatively stable presence, with slight variations depending on the specific facilitation task, indicating that their strengths may be more context-dependent. Claude, once again, appears with the smallest share, suggesting a more limited role in facilitation-based decision-making within this framework.

The balanced representation across all categories highlights the importance of model-specific facilitation strategies, as different AI systems

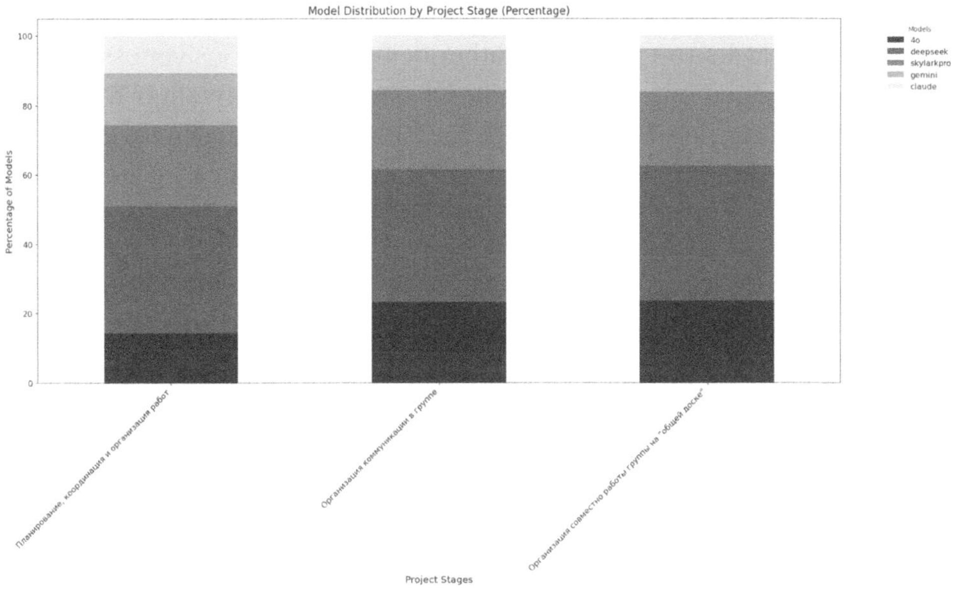

Figure 4.2 Model distribution by project stage.

demonstrate varying degrees of effectiveness depending on the specific nature of the project stage. This suggests that no single model universally outperforms the others across all facilitation scenarios, reinforcing the need for comparative evaluation methods such as the Swiss-style tournament employed in this study.

Conclusion and Future Directions

This study represents our initial endeavor into evaluating the potential of large language models (LLMs) as facilitators and moderators in strategic sessions. We have developed a preliminary evaluation framework focused on task analysis, benchmarking, solution development, implementation, and work planning. This framework draws inspiration from the Moscow Methodological Circle (MMC) and the system-thought-activity approach, which emphasize the intricate interplay between thought processes and activities [17].

The MMC tradition highlights more complex formats of collaborative intellectual activity, specifically "projecting" and "programming"—the latter pertaining to the design of project programs rather than computer programming. These formats necessitate more sophisticated facilitation norms, underscoring the need for advanced moderation strategies.

Our exploration has illuminated several pressing challenges: AI-generated facilitation often lacks the nuanced authenticity and responsiveness needed for sustained engagement; many existing tools do not fully align with established facilitation models that demand flexible, context-sensitive guidance; and both ethical dilemmas and technical barriers can limit user trust and broad adoption. Nonetheless, the theories of Freire, Vygotsky, Ilyenkov, and Dewey underscore the potential for AI to reinforce human-led facilitation when developers remain mindful of social equity, adaptive scaffolding, collective sense-making, and reflective inquiry. Such alignment with well-grounded educational practices encourages genuine dialogue, empowers participants to co-create objectives, and sustains iterative reevaluation. The resulting synergy between AI support and human facilitation can enrich decision-making processes, enhance learning outcomes in educational environments, and clarify organizational goals—provided developers, facilitators, and stakeholders collaborate to ensure transparent, ethically guided design and responsible implementation.

Future research directions include:

- **Expanding the Evaluation Framework**: We aim to enrich our database with new vignette situations and extend evaluations to encompass group projecting and programming modes.
- **Incorporating Human Evaluators**: Assessing AI-generated responses with human evaluators will provide nuanced insights into the effectiveness of LLMs in facilitation roles.
- **Refining System Prompts and Agentic Workflows**: Further experiments are planned to optimize system prompts and the agentic workflows of AI moderators, enhancing their adaptability and responsiveness.

By aligning AI facilitation tools with established educational practices that prioritize social equity, adaptive scaffolding, collective sense-making, and reflective inquiry, we can enhance decision-making processes, improve learning outcomes, and clarify organizational goals. This alignment requires collaborative efforts among developers, facilitators, and stakeholders to ensure transparent, ethically guided design and responsible implementation [10].

LLMs and other AI-driven systems hold considerable promise for transforming how small groups navigate strategic sessions, particularly as organizations face heightened complexity and rapid shifts in their environments. By automating routine tasks, offering real-time analysis of group dynamics, and weaving in adaptive interventions, AI can potentially ease the cognitive burden on human facilitators and amplify participant voices that might otherwise go unheard. Yet this exploration has also underlined several pressing challenges: AI-generated facilitation often lacks the nuanced authenticity and responsiveness needed for sustained engagement; many existing tools do not fully align with established facilitation models that demand flexible, context-sensitive guidance; and both ethical dilemmas and technical barriers can limit user trust and broad adoption. The resulting synergy between AI support and human facilitation can enrich decision-making processes, enhance learning outcomes in educational environments, and clarify organizational goals—provided developers, facilitators, and stakeholders collaborate to ensure transparent, ethically guided design and responsible implementation. Future research might refine AI's language generation capabilities, deepen its alignment with recognized facilitation frameworks, and embed more robust participatory mechanisms to build trust and cultivate psychological safety. In doing so, AI-driven facilitation can evolve from an experimental supplement to a valuable, human-centered companion for moderating small group intellectual efforts.

References

1. Argyris, C., & Schön, D. A. (1978). *Organizational learning: A theory of action perspective*. Reading, MA: Addison-Wesley.
2. Bens, I. (2017). *Facilitating with ease! Core skills for facilitators, team leaders and members, managers, consultants, and trainers* (4th ed.). San Francisco, CA: Jossey-Bass. https://www.perlego.com/book/996285/facilitating-with-ease-core-skills-for-facilitators-team-leaders-and-members-managers-consultants-and-trainers-pdf?
3. Biloslavo, R., Edgar, D., Aydın, E., et al. (2024). Artificial Intelligence (AI) and strategic planning process within VUCA environments: A research agenda and guidelines. *Management Decision*. Advance online publication. https://doi.org/10.1108/MD-10-2023-1944scispace.com
4. Bittner, E. A. C., Mirbabaie, M., & Morana, S. (2021). Digital facilitation assistance for collaborative, creative design processes. In *Proceedings of the 54th Hawaii international conference on system sciences* (pp. 370–379). https://doi.org/10.24251/HICSS.2021.044aisel.aisnet.org+3fis.uni-bamberg.de+3ResearchGate+3
5. Bittner, E. A. C., & Shoury, O. (2019). Designing automated facilitation for design thinking: A chatbot for supporting teams in the empathy map method. In *Proceedings of the 52nd Hawaii international conference on system sciences* (pp. 1–10). https://doi.org/10.24251/HICSS.2019.044

6. Chiang, C.-W., Lu, Z.-J., Li, Z., et al. (2024). Enhancing AI-assisted group decision making through LLM-powered devil's advocate. In *Proceedings of the 29th international conference on intelligent user interfaces* (pp. 1–10). https://doi.org/10.1145/3640543.3645199dl.acm.org+1dl.acm.org+1
7. Dewey, J. (1916). *Democracy and education: An introduction to the philosophy of education.* New York, NY: Macmillan.
8. Dong, Y., Ding, S., & Ito, T. (2024). An automated multi-phase facilitation agent based on LLM. *IEICE Transactions on Information and Systems*, E107-D(4), 1–9. https://doi.org/10.1587/transinf.2023IHP0011jstage.jst.go.jp+1jstage.jst.go.jp+1
9. Freire, P. (1970). *Pedagogy of the oppressed.* New York, NY: Continuum.
10. Vygotsky, L. S. (1978). *Mind in society: The development of higher psychological processes.* Cambridge, MA: Harvard University Press.
11. Edmondson, A. C. (1999). Psychological safety and learning behavior in work teams. *Administrative Science Quarterly*, 44(2), 350–383. https://doi.org/10.2307/2666999scirp.org+1dash.harvard.edu+1
12. Houtti, M., Zhou, M., Terveen, L., & Chancellor, S. (2025). Observe, ask, intervene: Designing AI agents for more inclusive meetings. In *Proceedings of the 2025 CHI conference on human factors in computing systems* (pp. 1–29). https://doi.org/10.1145/3706598.3713838arXiv+1dl.acm.org+1
13. Ilyenkov, E. V. (1977). *Dialectics of the ideal.* Moscow: Progress Publishers.
14. Zheng, L., Long, M., Niu, J., & Zhong, L. (2023). An automated group learning engagement analysis and feedback approach to promoting collaborative knowledge building, group performance, and socially shared regulation in CSCL. *International Journal of Computer Supported Collaborative Learning*, 18(1), 101–133.
15. Liu, X. (2023). Opportunities of artificial intelligence for supporting complex problem-solving: Findings from a scoping review. *Computers & Education: Artificial Intelligence*, 4, 100138. https://doi.org/10.1016/j.caeai.2023.100138ouci.dntb.gov.ua
16. Morgan, G. (2006). *Images of organization* (Updated ed.). Thousand Oaks, CA: Sage Publications.
17. Park, S., Nixon, N., Doroudi, S., et al. (2024). Advancing knowledge together: Integrating large language model-based conversational AI in small group collaborative learning. In *Proceedings of the 2024 CHI conference on human factors in computing systems* (pp. 1–10). https://doi.org/10.1145/3544548.3581132
18. McGregor, M. (2025, January 7). How teachers can orchestrate a classroom filled with AI tools. *Getting Smart.* Retrieved from https://www.gettingsmart.com/2025/01/07/how-teachers-can-orchestrate-a-classroom-filled-with-ai-tools/
19. Nixon, N., Lin, Y., & Snow, L. (2024). Catalyzing equity in STEM teams: Harnessing generative AI for inclusion and diversity. *Policy Insights from the Behavioral and Brain Sciences*, 11(1), 1–10. https://doi.org/10.1177/23727322231222809

Chapter 5

Multimodal Embeddings for Predicting States of Negotiators, Their Changes, and Contents

Vladislav Zapylikhin, Alexander Didenko,
Mikayil Majidov, and Yana Milenkaya

Introduction

Recent developments in the analysis of verbal and nonverbal signals open up new opportunities for an objective investigation of the dynamics of interpersonal interactions, particularly in the context of business negotiations. The relevance of this research is driven by the growing popularity of applying machine learning in the social domain, the advancement of computer vision technologies, and the increasing number of studies in these areas. The negotiation process, characterized by constant changes in participants' states, directly influences the outcome of deals, which underscores the need to develop automated tools for their analysis.

The aim of this study is to identify a discrete set of human states, determine their semantic basis, and accurately predict the transitions between negotiators' states using machine learning methods. Special emphasis is placed on applying neural network algorithms to integrate both verbal and

DOI: 10.4324/9781003606956-5

nonverbal data, thereby providing a more comprehensive understanding of the dynamics of the negotiation process.

This research addresses the following questions:

- Is the influence of a participant's state, as exhibited during paired business negotiations, on the state of their opponent predictable?
- Are neural network algorithms applicable for effectively determining negotiators' states and predicting their transitions?

Based on these questions, the following hypotheses are proposed:

Hypothesis 1 posits that neural networks can extract a discrete number of clusters representing negotiators' states based on both their emotional reactions and verbal cues. The effectiveness of these clusters will be evaluated using metrics such as the Silhouette Score, Davies-Bouldin Index, Dunn Index, *F*-statistic, and *Z*-statistic.

Hypothesis 2 asserts that machine learning algorithms can accurately predict transitions between negotiators' states using the temporal dynamics present in multimodal data, as measured by accuracy, precision, recall, and ROC AUC.

Hypothesis 3 suggests that multimodal embeddings that combine both verbal and emotional information provide a stronger predictive basis for state transitions than features derived solely from spoken words, as determined by feature importance analysis. The outcomes of these evaluations will support the integration of verbal and emotional data for robust modeling of negotiators' state transitions.

Thus, this research is aimed at developing tools for analyzing the dynamic nature of negotiators' states, which is crucial for optimizing negotiation processes in the modern digital economy.

Related Works

Negotiation analysis attracts the attention of researchers from various disciplines, including linguistics, psychology, and artificial intelligence. This section reviews key studies organized into five thematic areas: analysis of speech in negotiations, the role of emotions in negotiations, interaction and communication between individuals, theories and schools of negotiation, and methods for obtaining multimodal embeddings.

Speech in Negotiations

Speech plays a central role in negotiations by shaping participants' perceptions and ultimately influencing the outcome of interactions. Bern's transactional analysis [1] posits that communication occurs through the interplay of three ego states—Parent, Adult, and Child—demonstrating that speech patterns may reveal internal psychological states. From a cognitive standpoint, Vygotsky [2] argued that language not only expresses thought but also shapes cognitive processes and decision-making. In discursive theory, Laclau and Mouffe [3] asserted that language does not simply describe reality but actively constructs it—a perspective that aligns with Foucault's view that discourses embody power structures which delineate the boundaries of interpersonal communication [4].

Emotions in Negotiations

Emotions exert a significant influence on the negotiation process by affecting participants' behavior and their willingness to compromise. Darwin [5] and Ekman [6] contended that human facial expressions convey a universal set of basic emotions, an idea that underpins current facial emotion recognition (FER) systems. In contrast, Plamper [7] questions the universality of emotions, suggesting instead that they are social constructs that vary according to cultural context. Moreover, Russell's circumplex model [8] conceptualizes emotions as existing on a continuous spectrum rather than as discrete states.

Interaction and Communication in Negotiations

The social dimensions of negotiations have been explored through various psychological and sociological theories. Altman and Taylor [9] described the deepening of interpersonal relationships through gradual self-disclosure in their social penetration theory, while Bruner [10] identified mechanisms of interpersonal perception—such as identification, empathy, and causal attribution—that shape negotiators' behavior. Research by Tiedens and Fragale

[11] demonstrates that dominant behavior can influence negotiation outcomes, and Goffman [12] introduced the notion of social roles, proposing that individuals act in ways that conform to the expectations of others.

Negotiation and Major Negotiation Schools

Different schools of negotiation offer distinct interpretations of strategies and methods for conducting dialogue. Walton and McKersie [13] differentiated between distributive (competitive) and integrative (cooperative) negotiation, whereas Fisher and Ury [14] proposed a principled negotiation method centered on achieving mutually beneficial agreements. A cross-cultural perspective presented by Hofstede [15] further illustrates how cultural differences influence negotiation styles, while Bazerman and Neil [16] have highlighted cognitive biases that impede rational decision-making. Most existing models view negotiations as static processes that fail to account for the dynamic nature of negotiators' states; the present study addresses this gap by applying sequential state modeling based on multimodal embeddings.

Multimodal Emotion Recognition

Multimodal emotion recognition systems in general aim to capture complementary emotional cues from multiple modalities (visual, acoustic, physiological, and text) to enhance robustness and accuracy. One possible approach here is to find a way of projecting features from multiple modalities to shared feature space of multimodal embeddings.

Early works in this flow of literature relied on handcrafted features and rule-based systems, but the advent of deep learning has shifted this emphasis to data-driven methods. Subsequent models have evolved to integrate modality-specific feature extractors (e.g., CNNs for images, RNNs for speech) which are then combined through either explicit fusion (discussed in the next paragraph) or alignment into shared representations.

The primary challenge in MER is to manage the inherent heterogeneity among modalities—differences in temporal dynamics, noise, and semantic granularity—which has motivated the development of specialized integration techniques [17]. Recent work has shown that joint embeddings derived from multiple modalities could yield richer affective representations. These embeddings enable robust emotion discrimination by capturing cross-modal semantic structures. Unlike traditional fusion, embeddings-based methods emphasize learning a unified latent space where multimodal data aligns effectively. The primary challenge lies in mitigating heterogeneity while maintaining the complementarity of modality-specific information.

Signal Fusion for Emotion Recognition

Fusion-based approaches focus on directly combining information from multiple modalities at various stages of processing. Unlike embedding-based methods, fusion emphasizes the direct interaction between modalities rather than aligning them into a shared space. This can be categorized as Early, Intermediate, and Late Fusion, on which we elaborate below.

Early Fusion

Early Fusion concatenates raw or low-level features from each modality, enabling joint processing at the outset. This approach is highlighted in several studies, demonstrating its effectiveness in capturing complex interactions between modalities such as text, audio, and visual data. For instance, Lee et al. found that early stage fusion significantly outperformed late stage fusion in a transformer-based architecture for sentiment analysis, achieving an accuracy of 71.87% on the CMU-MOSEI dataset. Similarly, Ge et al. employed an early fusion strategy to jointly train audio and text modalities, which helped mitigate modal competition and improved performance in the Multimodal Emotion Recognition Challenge (MER2024). Lei's study further supports the efficacy of feature-level fusion, which amalgamates data types early in the process, thereby improving detection robustness and surpassing single-modal systems. The survey by Mohamed et al. also underscores the potential of early fusion, particularly when combined with context-aware approaches, to enhance affective state recognition in human-robot interaction scenarios [18].

Intermediate Fusion

This approach involves integrating features from different modalities at a stage between initial feature extraction and final decision-making, allowing for more nuanced interactions between modalities. Usually approaches of this type extract modality-specific representations via deep encoder networks, and then integrate them using techniques such as cross-attention ([19, 20] or gating mechanisms [22] and other). Conceptually, the difference between the two is that cross-attention allows one modality to extract relevant information from another by computing Query-Key-Value mappings, thereby modeling inter-modal dependencies through learned attention weights. In contrast, gating mechanisms dynamically regulate how much each modality contributes by scaling input features with non-linear activation functions (e.g., sigmoid), which helps suppress noisy or redundant information. Conversely, gating mechanisms do not explicitly model dependencies between modalities but instead prioritize the most reliable signal. Hence the difference in tasks where these two techniques are usually used: while gating is particularly beneficial in scenarios where data quality varies across modalities, cross-attention is more effective for tasks requiring intricate semantic correspondence, such as mapping audio signals to textual content. For instance, the work by Ming Xu et al. introduces a hierarchical cross-modal spatial fusion network that integrates EEG and video features, employing a coordinated attention module to enhance interaction between modalities, achieving notable accuracy on datasets like DEAP and MAHNOB-HCI. Similarly, the study by R. Gnana Praveen and Jahangir Alam presents a Recursive Joint Cross-Modal Attention (RJCMA) mechanism that captures intra- and inter-modal relationships across audio, visual, and text modalities, significantly improving performance on the Affwild2 dataset [20]. Another approach is the intermediate-layer attention mechanism proposed by P. Sreevidya and Sathishkumar Samiappan, which uses inter- and intra-attention mechanisms for fusion, further refined with a binary search algorithm to enhance feature discernment for emotion recognition tasks [21]. Further, the Gated Attention Fusion Module, as part of a cascaded encoder-decoder reconstruction network, is designed to handle missing modalities and multimodal imbalance, effectively learning robust representations under different conditions. Another approach, the Multimodal Recurrent Intermediate-Layer Aggregation (RILA) model, leverages the multimodal aggregation gating module to capture utterance-level interactions across modalities and layers, enhancing the performance of emotion recognition by fully exploiting intermediate representations [23]. The Group Gated Fusion (GGF) layer in the Gated Bidirectional Alignment Network (GBAN) integrates representations from different modalities, showing significant improvements over traditional methods by focusing on temporal alignment and attention-based bidirectional alignment. Additionally, the Recursive Joint Cross-Modal Attention (RJCMA) model captures intra- and inter-modal relationships through recursive mechanisms, further refining feature representations and improving emotion recognition accuracy [20]. The VISTA Net employs a hybrid fusion approach, adjusting weights of intermediate outputs automatically, which enhances interpretability and accuracy in emotion classification [22]. These studies collectively highlight the importance of intermediate fusion gating mechanisms in addressing challenges such as missing modalities, synchronization, and feature extraction, ultimately leading to more accurate and reliable multimodal emotion recognition systems.

Late Fusion

Late fusion defers the integration process until after each modality has been processed independently. In late fusion, high-level decisions or probability distributions from separately trained unimodal models are combined at the final stage, ensuring that each modality contributes without direct feature-level interactions during training. This approach reduces the risk of inter-modal interference and is particularly useful when modalities have vastly different data distributions, temporal structures, or noise characteristics. This approach is highlighted in several studies, and many of them were shown as a performance booster for emotion recognition models. For instance, Georgescu et al. integrates speech and visual modality, achieving a notable accuracy of 97.95% in their multimodal system, which surpassed individual modality performances [24]. Similarly, SriHari and Bhaskaran implemented a simple late fusion model architecture that combined speech emotion recognition and facial emotion expression methods, resulting in a 97.2% accuracy on the RAVDESS dataset [25]. Zhu's review also emphasizes decision-level fusion, which consolidates outcomes from different modalities to ensure stable classification, highlighting its utility in multimodal emotion analysis [26]. Furthermore, Ibrahim et al. demonstrated improved performance in multimodal emotion recognition through late fusion and other techniques, outperforming unimodal approaches [27]. In contrast, Wang et al. [28] employed decision-level fusion to integrate three modalities, achieving both high accuracy and strong generalization in emotion recognition. However, late fusion lacks the ability to refine multimodal representations at earlier stages, as seen in intermediate fusion methods, which leverage cross-attention for inter-modal feature alignment and gating for adaptive weighting. Consequently, while late fusion offers modularity and robustness in scenarios where direct modality interactions may introduce noise, it may not capture the intricate cross-modal relationships that intermediate fusion techniques exploit.

Graph-Based Modeling

Graph-based approaches represent modality data as nodes within a graph, the edges encoding relationships or dependencies. This method differs fundamentally from traditional fusion and embeddings because it focuses on the relational structure of data rather than direct feature aggregation. By modeling higher-order interdependencies—such as spatial facial feature arrangements or temporal correlations—graph-based models dynamically adapt to evolving emotional patterns, making them uniquely suited for capturing contextual dependencies. Unlike fusion techniques, which impose a predefined integration strategy, and embeddings aiming for homogeneous multimodal representations, graph approaches try to preserve modality-specific structures and interactions, allowing for more flexible, context-aware processing. For example, in affective computing, physiological signals (e.g., EEG, heart rate) and facial expressions often exhibit correlated patterns—e.g., stress leading to both increased heart rate and furrowed brows. A graph-based model can preserve cross-modal structural dependencies by linking specific facial regions (e.g., forehead tension) with physiological nodes (e.g., increased skin conductance). An embedding model, however, may force both signals into a uniform space, losing the fine-grained physiological-visual mapping.

Graph-based modeling in multimodal emotion recognition has been explored through various approaches, as highlighted in the provided papers. The Capsule Graph Transformer Architecture proposed by Filali et al. utilizes a graph convolutional network to integrate multimodal inputs, such as text, speech, and visual cues, achieving state-of-the-art results on datasets like MELD and MOSEI [29]. Bhattacharyya et al. introduced the HMG-Emo framework, which leverages heterogeneous graph learning for personalized emotion prediction in social networks, demonstrating the effectiveness of graph neural networks in outperforming traditional methods [30]. The framework introduces a heterogeneous graph to model interactions in social networks. This graph consists of nodes representing users and images, and edges representing interactions between users and images (e.g., viewing an image) and connections between users (e.g., social ties). The graph structure allows for the integration of various contextual information, such as user comments and group memberships, as node and edge features

Shi et al. proposed a multimodal graph learning framework using framelet-based stochastic configuration networks, which addresses the limitations of existing graph-based models by capturing both low- and high-pass components to improve emotion recognition in conversations [31]. Similarly, Feng and Fan developed a method incorporating adaptive graph convolutional networks to construct dialogue relationship graphs, enhancing the extraction of dependencies between speakers and achieving high recognition accuracy [32]. Zhao et al. introduced a bilevel relational graph approach for emotion recognition in conversation, which constructs global affinity and local context dependency graphs to preserve discriminative information and efficiently extract long-range contextual information [33]. These studies collectively underscore the potential of graph-based models in enhancing the accuracy and robustness of multimodal emotion recognition systems by effectively capturing complex relationships and interactions across different modalities.

Multiple Instance Learning (MIL) Approaches

MIL frameworks reinterpret modality data as sets of instances, identifying the most informative segments for emotion recognition. Unlike traditional fusion or embedding approaches, which assume all modality features contribute equally, MIL dynamically weighs the relevance of each instance, filtering out noise and irrelevant details. This focus on instance-level relevance is particularly useful for long-duration videos or physiological monitoring, where emotions evolve over time [34].

Teacher–Student and Distillation-Based Approaches

Teacher–student paradigms transfer knowledge from complex multimodal teacher models to efficient student models via knowledge distillation. Unlike fusion or embedding techniques, which require explicit integration of multimodal signals at runtime, teacher-student frameworks emphasize knowledge transfer. This enables deployment of lightweight models that retain multimodal awareness without requiring direct fusion during inference.

Temporal and Sequential Modeling with Memory Mechanisms

Sequential modeling techniques, such as LSTMs, transformers, and gated recurrent units, capture temporal dependencies within each modality. Unlike static fusion or embedding methods, which treat modalities independently or align them into a single representation, temporal modeling dynamically integrates past and present emotional cues. This makes them particularly suited for applications requiring context-dependent understanding, such as conversational emotion recognition.

Multi-Task Learning and Contextual Approaches

Multi-task learning (MTL) integrates emotion recognition with auxiliary tasks like sentiment analysis or speaker identification. Unlike fusion-based approaches that explicitly combine features, MTL frameworks share representations implicitly across tasks. This enables richer contextual understanding without requiring strict modality alignment, making them distinct from other integration techniques.

Methods for Obtaining Multimodal Embeddings

Recent advances in machine learning enable the fusion of data from different modalities into a unified vector space. The CoLES approach [35] employs contrastive learning for sequential data, effectively capturing temporal dependencies, while the Replaced Token Detection (RTD) method [36] enhances contextual representations in text via optimized pre-training. Other approaches, such as ImageBind [37] and Meta-Transformer [38], integrate text, audio, and video into a single representation. However, most prior studies have focused on retrieval tasks rather than on predicting changes in states. In this work, multimodal embeddings are applied to dynamically model negotiators' states.

Methods

This section establishes the methodological rigor underpinning our analysis of multimodal emotion dynamics in negotiation contexts, addressing both empirical design and computational challenges unique to dyadic interaction data. The subsequent subsections systematically detail the experimental protocol, data synchronization procedures, and neural architecture selection criteria—structured to ensure reproducibility while mitigating real-world constraints such as temporal misalignment and cross-modal noise.

Experimental Setup, Sample, and Data Collection

The empirical basis of the research consisted of recordings of 67 dyadic negotiations collected by the research group. All subjects—students from various courses in the "Management" specialty—were selected from three cities: Moscow, Krasnogorsk, and Kaliningrad, in Russia.

The experimental procedure was carried out as follows:

Briefing and distribution of materials. Printed materials containing a description of a conflict situation were distributed to the students (Figure 5.1). At the same time, the participants were informed that the initial facts in the scenario should not be changed; however, during the negotiations, they were allowed to add new input data, which should be accepted by both parties as true. Any input to be introduced further should be compatible with previous inputs. It was prohibited to introduce inputs that would arbitrarily resolve the conflict without logical consistency or contextual justification, in the "deus ex machina" style.

Equipment and arrangement: Each participant was provided with an individual microphone. Two cameras were placed on the table, each of which was directed toward the corresponding participant. The subjects held the printed materials describing the situation throughout the negotiations.

Negotiation procedure: All equipment was started sequentially, and at the signal, the participants were informed of the beginning of the negotiations. The duration of the negotiations was limited to 5 minutes. One minute before the end, a sound signal was given to the participants, after which, upon expiration of the allotted time, a signal was given indicating the end of the negotiations. Before the start of the negotiations, the participants chose one of three roles: director, investor, or director's friend ("Pyotr"). Some students participated in the experiment twice; however, the selection of a previously played role was prohibited for repetition.

```
Situation "Share"

When creating a new company, three friends offered the Investor 51% of the
company. The three friends received 24% and the positions of director and
two deputies. They were to register the other 25% in their name in two
years, provided that as a result of their work certain results would be
achieved: the company would occupy a certain market share, and additional
funds would be attracted in the form of loans.

The company has been developing steadily for two years, but the results that
were agreed upon have not been achieved. The market share is occupied twice
as small, loans have not been attracted. Development is only happening with
the Investor's money.

Since the time has come to register 25% of the shares, the director and his
friends decide to ask the Investor to register 25% of the shares in their
name. The Investor, taking advantage of the fact that results have not been
achieved, wants to sell 25% of the shares to a third-party company to
attract working capital. One of the friends, Peter, supports the investor,
since he considers his position to be legitimate.

ROLES AND INTERESTS:

Investor - interested in developing the company and attracting additional
working capital. Wants to sell shares to the side. At the same time, wants
to maintain relations with the director, because he does not see an
alternative to him yet.

Director - wants to register 25% of the shares in his own name and that of
his friends, at the same time wants to maintain relations with the Investor.

Pyotr - does not want a quarrel with the Investor.
```

Figure 5.1 Example of an input for dyadic negotiation.

Sample characteristics: The total number of participants was 134 individuals (75 males and 59 females). The distribution by roles and by gender varied depending on the city, as shown in Table 5.1.

Speech and Emotion Recognition, as well as Their Synchronization

During the experiment, the following issues arose:

- Unsynchronized device startup. The cameras and microphones were activated sequentially—first for one participant, then for the other, which led to differences in the initial time stamps. In addition, the

Table 5.1 Distribution of Players by Role

Kaliningrad	*Male*	*Female*
Investor	9	16
Director	13	14
Director's friend	3	1
Krasnogorsk	*Male*	*Female*
Investor	18	10
Director	17	10
Director's friend	4	1
Moscow	*Male*	*Female*
Investor	4	3
Director	6	2
Director's friend	1	2

audio recording was carried out with a separate device, which further complicated the synchronization.

- Positioning problems. Some participants approached the camera too closely or covered their face with their hands, which negatively affected the quality of emotion recognition.
- Cross-interference noises. Due to the simultaneous conduct of negotiations, the microphones sometimes captured not only the target participant's speech but also that of their opponent or bystanders.

In view of the aforementioned issues, the synchronization of the audio and video recordings was carried out predominantly manually. A developed program allowed for automatic trimming of the recordings according to specified time stamps (for example, by the signals indicating the beginning and end of the negotiations); however, the final adjustment was performed by an operator.

To extract textual data from the audio recordings, the Whisper model [39] in its large configuration was employed, offering high speech recognition accuracy and automatic punctuation insertion, which enhances the quality of subsequent text analysis. The emotional states of participants were assessed using the facial emotion recognition (FER) library through the analysis of video recordings, where facial expression changes were processed for each utterance, yielding a probability vector corresponding to seven emotions: anger, disgust, fear, joy, neutral state, sadness, and surprise.

Analysis of the Features of the Negotiations

To obtain initial representations of the data, the following stages of analysis were conducted: Analysis of the duration of utterances. A distribution graph was constructed of the durations of individual phrases, which had the form of a logarithmically normal distribution with a slight lag (the duration of individual phrases exceeded 40 seconds), which can be seen in Figure 5.2.

The analysis of general speech activity is shown in Figure 5.3. The distribution of the total speaking time of the negotiators was examined (for example, 40% of the participants spoke for 2–3 minutes, 6% for up to 4.1 minutes, and 6% for less than 1 minute).

Analyses of facial expressions are shown in Figures 5.4 and 5.5. The distribution of the frequency and duration of expressed emotions was analyzed for each participant. It was noted that, on average, there were about 895 changes in facial expressions per utterance, and most of the time the participants remained in a neutral state.

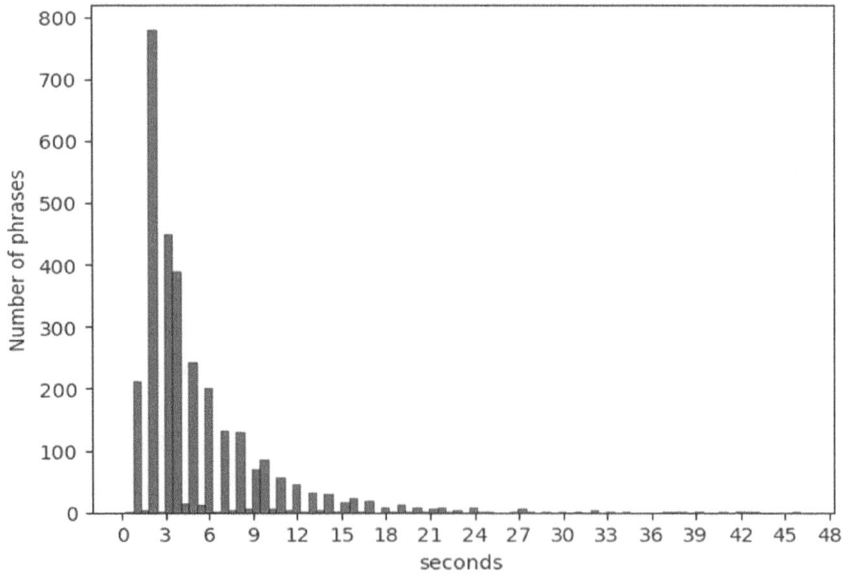

Figure 5.2 Distribution of the duration of a single utterance.

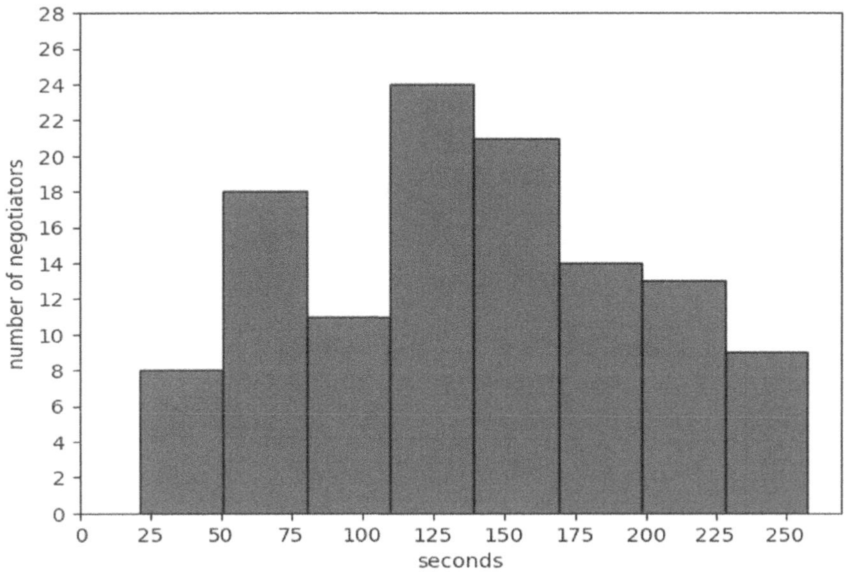

Figure 5.3 Distribution of the total speaking time of the negotiators.

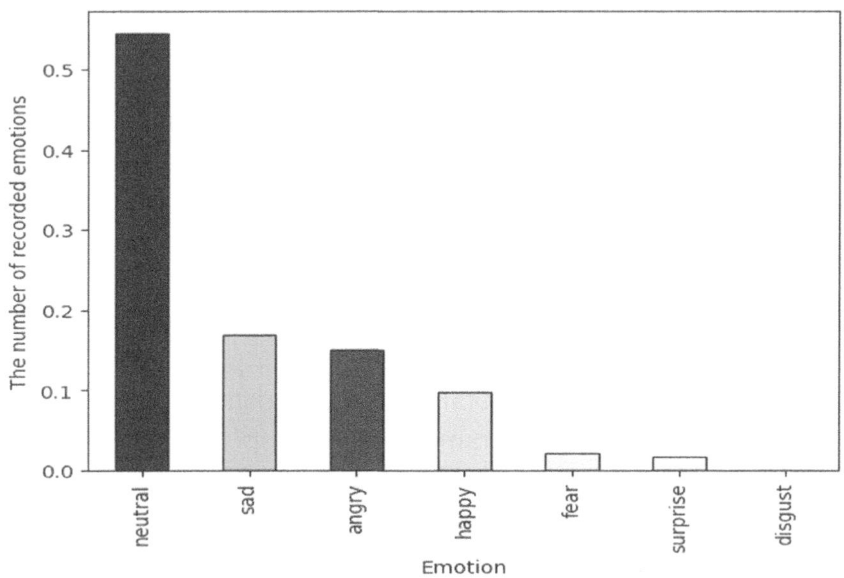

Figure 5.4 Frequency of expressed emotions.

Selection and Analysis of Neural Network Models for Creating Multimodal Embeddings

Since the negotiation data represent complex time series consisting of textual (speech) and numerical information (emotion vectors), an analysis of

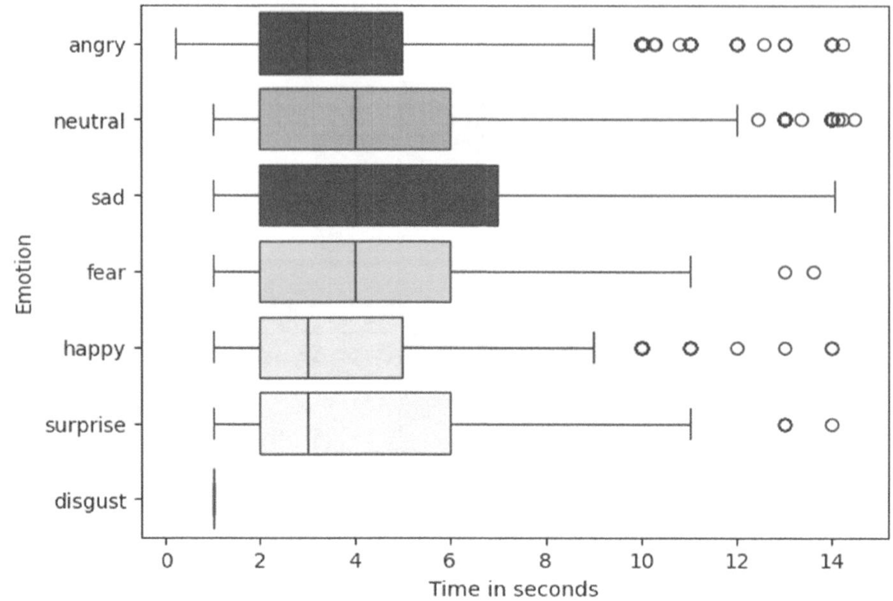

Figure 5.5 Duration of expressed emotions.

existing frameworks for creating multimodal embeddings was conducted. The following libraries were considered:

ImageBind. This library provides the capability to obtain embeddings based on six modalities (images, text, audio, depth, thermal map, IMU); however, due to limited support for combinations of data and the incompleteness of the implementation, this library was rejected.

Meta-Transformer. A framework that uses a unified tokenizer and a fixed encoder to extract high-level semantic characteristics from various modalities. Despite its wide capabilities, Meta-Transformer proved to be less suitable for the specifics of our temporal negotiation data.

PyTorch LifeStream (PTLS). A framework developed by the Sber AI Laboratory, intended for working with complex structured sequential data. It allows for obtaining embeddings that take into account the context and relationships between different modalities. In PTLS, the following unsupervised learning methods are implemented:

- Contrastive Predictive Coding (CPC)
- Replaced Token Detection (RTD)—a method first proposed in the paper "ELECTRA"
- Pre-training Text Encoders as Discriminators Rather Than Generators [36]

- Next Sequence Prediction (NSP) from BERT
- Sequences Order Prediction (SOP) from ALBERT
- Masked Language Model (MLM) from RoBERTa
- Contrastive Learning for Event Sequences (CoLES) – a method proposed for working with event data [35].

Of all the proposed methods for obtaining multimodal embeddings, we selected RTD and CoLES, as conceptually they are best suited for our task. The principle of contrastive learning consists in minimizing the loss function for similar objects and maximizing it for dissimilar ones, which ensures effective separation of embeddings

Obtaining Clusters of Negotiators' States and Predicting State Transitions

After obtaining embeddings for each utterance (embedding dimension—1×256) using the module based on RTD and CoLES, the stage of dimensionality reduction and clustering was carried out:

Dimensionality reduction. The t-SNE (t-distributed Stochastic Neighbor Embedding) method was applied for visualizing high-dimensional data in a two-dimensional space.

Additionally, the PCA (Principal Component Analysis) method was used, which allows finding linear combinations of features while preserving maximum variance, after which a clustering method was applied.

Clustering. For grouping the embeddings, the K-means algorithm was used, which divides the data into a specified number of clusters, minimizing the total distance between the points and their centroids.

Construction of the Predictive Model. To test the hypotheses regarding the predictability of transitions in negotiators' states, a machine learning model was created based on the gradient boosting algorithm using the scikit-learn library.

Target variable: a binary label, where 1 indicates a change in the negotiator's state and 0 indicates no change.

Observed features:

- Negotiation identifier
- Multimodal embeddings obtained from utterances and emotional data

- The number of the cluster to which the current state belongs
- Semantic features of the text extracted based on the annotation of utterances using a dictionary of evaluative words and expressions (e.g., RuSentiLex)

To evaluate the effectiveness of the model, the following metrics were used: accuracy, precision, recall, and ROC AUC. Additionally, to test the hypothesis regarding the possibility of distinguishing discrete clusters of states, the following measures were applied: Silhouette Score, Davies–Bouldin Index, Dunn Index, F-statistic, and Z-statistic for means.

Thus, the described methodology includes a comprehensive approach to the collection, synchronization, and processing of negotiation data, the generation of multimodal embeddings using the RTD and CoLES methods, as well as subsequent clustering and the construction of a predictive model.

Results

As a result of the clustering analysis, five clusters were identified: two clusters with nearly equal numbers of points, two less distinct and less populated clusters, and one clearly defined separate cluster. The primary visual requirement for the resulting clusters was that the grouping of points should be based on a principle different from the obvious grouping criteria of the original data (i.e., grouping by dialogues, emotions, or roles). All clusters turned out to be composed of utterances from different parts of the majority of dialogues, with a variety of emotions present and no grouping by roles. In general, the clusters represent segments of consecutive utterances from both participants in certain parts of the dialogues. Here, we present a detailed examination of the largest clusters, which cover 85.9% of the points.

Let us consider Cluster "0" in more detail, which contains 1388 points collected from 89.7% of the conducted dialogues. An example from cluster 0 is shown in Table 5.2. In most of the utterances in this cluster, the participants express confidence in their words, a certain assertiveness, and a lack of conciliation. Emotionally, the participants exhibit neutrality, anger, sadness, and occasionally joy.

Cluster "3" contains 1243 points covering 88.2% of the dialogues. An example from cluster 0 is shown in Table 5.3. Based on the analysis, it can be concluded that the majority of the utterances in this cluster are united by the participants' expression of uncertainty. Additionally, the participants

Table 5.2 Cluster "0"

Cluster No.	*Negotiation No.*	*Negotiator's Utterance*	*Emotion*	*Semantic Group of Similar Utterances*
0	3	"We have not secured any loans, so I propose to lower the interest rate on loans so that you take more loans from us."	Angry	Indication of dissatisfaction
0	9	"Our enterprise is only two years old. You must understand that at the early stages, a lot of investment is needed."	Neutral	Assertiveness
0	16	"Well, why should I give you a second chance?"	Sad	Indication of dissatisfaction
0	33	"In two years, we will show you the dynamics, you will see the figures, and you must understand that no one else can do it better than us."	Neutral	Demonstration of confidence

more frequently use various words that express their agreement with the opponent's arguments. Emotionally, there is a predominance of sadness and a neutral state.

Cluster "4" is relatively small, containing 124 points. In this cluster, participants discuss the format in which the contract can be renegotiated and give promises. Overall, a willingness to converse is demonstrated, and a calm atmosphere is present. The dominant state of the participants is neutral, with anger and sadness occurring rarely.

Cluster "1" is the smallest by the number of points, comprising 62 points. In this cluster, nearly all points represent a neutral state of the participants; their utterances suggest that the participants are somewhat passive in the discussion, avoid accusations, more often state facts, and then hand over the floor to their opponent.

Table 5.3 Cluster "3"

Cluster No.	*Negotiation No.*	*Negotiator's Utterance*	*Emotion*	*Semantic Group of Similar Utterances*
3	51	"Well, the proposal is certainly interesting, but how can I trust you again when you have already let me down once, with no guarantee that in these same three months you won't tell me the same thing—that you once again failed to meet the deadline?"	Neutral	Uncertainty
3	16	"Well, are you sure that within this period, I would not suggest an alternative—namely, that if you do not meet the deadline, you sell the shares to other people?"	Sad	Uncertainty
3	35	"Well, I completely agree with you. The 25% is ours."	Neutral	Agreement

Cluster "2" contains 244 points. The participants' argumentation in this cluster can be described as rather weak; they mainly provide not facts but rather vague reflections on the company's failures. The language used by the participants in this cluster cannot be described as businesslike, and the uttered phrases are quite conversational.

To test Hypothesis 1—that neural networks can extract a discrete set of negotiator state clusters based on their emotional expressions and spoken words—we computed a suite of clustering quality metrics, including the Silhouette Score, Davies-Bouldin Index, Dunn Index, *F*-statistic, and *Z*-statistic for the means.

Interpreting the obtained metrics, the Silhouette Score was 0.32. In general, Silhouette Score values can range from –1 to 1, where a value close to 1 indicates that the data points are well grouped within their clusters, a value close to 0 indicates that the data points are on the boundary between clusters, and a value close to –1 indicates that the data points may have been incorrectly clustered. In our case, the metric indicates that the clusters were obtained overall correctly, although a sufficient number of points lie on the cluster boundaries. For the Davies-Bouldin Index, values between 0 and 1 are generally considered good, especially if they are significantly below 1. This indicates that the clusters are compactly separated. Values above 1 may indicate poorer clustering, where clusters are less compact and/or poorly separated. In our case, the value was 0.69, which is considered a good level of data clustering. The Dunn Index evaluates the quality of the clustering by measuring the minimum distance between clusters and the maximum intra-cluster distance. High Dunn Index values indicate good clustering, while low values indicate poor clustering. For our data, the Dunn Index is 0.54, which is low; therefore, by this criterion, our clustering is considered insufficiently effective. The *F*-statistic yielded quite high values (*F*-ratio for feature 0: 2437.79, *F*-ratio for feature 1: 1090.44), meaning that the inter-cluster variance (differences between clusters) significantly exceeds the intra-cluster variance (differences within clusters). This indicates that feature 0 and feature 1 effectively separate the clusters. The results of the *Z*-test are presented in Table 5.4.

The *Z*-test results for the mean values between clusters show significant differences between the cluster centroids. The only exception is the pair of clusters 0 and 1 for feature 1, where the differences are not statistically significant. Evaluating these results, it can be concluded that Hypothesis 1—stating that neural networks can be used to obtain a discrete number of clusters of negotiators' states based on data about their emotions and the words they utter—is confirmed.

For testing Hypothesis 2, the metrics accuracy, recall, precision, and ROC AUC were calculated. The results of these metrics are shown

The ROC AUC score (0.66) indicates that the model has a moderate ability to distinguish between positive and negative classes—better than random guessing, but there is room for improvement. The accuracy shows that the model correctly predicts 70% of all observations. The recall indicates that the algorithm correctly identifies 70% of all truly positive cases, but misses 30%. Precision can be interpreted such that the model correctly predicts 71% of all positive cases, while 29% of the predicted positive cases are false. For data such as emotions and text, the model can be said to predict state transitions reasonably well, though there is potential for further improvement. This implies that neural network models can predict changes in negotiators' states.

For testing Hypothesis 3, the method feature_importances_ from the scikit-learn library was applied to the observed features. The

Table 5.4 *Z*-Test in Testing Hypothesis 1

Index	*Cluster 1*	*Cluster 2*	*Z-Test for Feature 0*	*Z-Test for Feature 1*	P-*Value for Feature 0*	P-*Value for Feature 1*
0	0	1	−140.467804	−0.887412	0.0	0.374857
1	0	2	−151.489944	40.106804	0.0	0.000
2	0	3	−52.510963	56.153267	0.0	0.000
3	0	4	−137.395279	−47.369827	0.0	0.000
4	1	2	−10.258543	38.636154	0.0	0.000
5	1	3	73.367508	52.931717	0.0	0.000
6	1	4	37.983910	−40.313087	0.0	0.000
7	2	3	82.355865	8.594214	0.0	0.000
8	2	4	55.194641	−88.698700	0.0	0.000
9	3	4	−60.007992	−128.458710	0.0	0.000

feature_importances_ method is used to evaluate the importance of each feature in the model. It is available for decision trees and ensemble methods such as RandomForest and GradientBoosting. The importance of a feature is determined as the sum of the decreases in the criterion (e.g., Gini impurity or entropy) at all nodes where that feature is used for splitting, averaged over all trees in the ensemble. It should be recalled that, in addition to the embeddings, text-based features were added: part of speech, semantic polarity (positive, negative, or neutral), as well as the source (whether the word is an evaluation, an opinion, or a fact). The results are presented in Figure 5.6.

As can be seen, the embedding and cluster features are more significant than the text features. The text features did not appear among the top 20 most significant features. This not only demonstrates that features based on text and emotions are more important for predicting state transitions, but also reflects the fact that a person's state—in our case, that of the negotiator—cannot be determined solely by the words they utter. The state is much more complex and encompasses both the spoken text and the speaker's emotions.

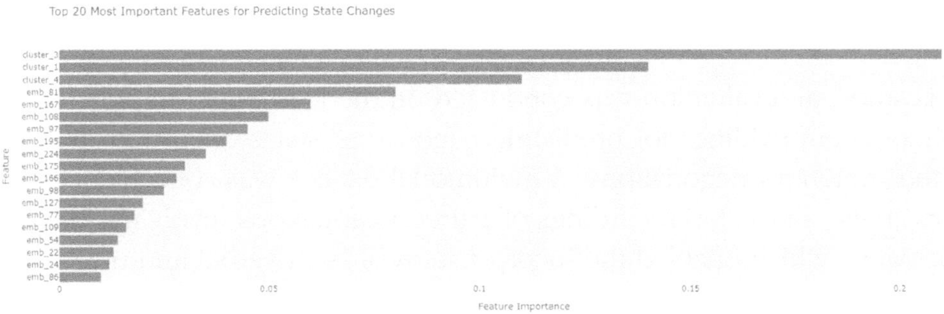

Figure 5.6 The 20 most significant features for predicting state transitions.

Discussion

In this study, we successfully compiled a unique multimodal dataset, comprising audio and video recordings of paired negotiations, which allowed us to account for both the verbal and nonverbal aspects of communication. The obtained results confirm that neural network models can effectively extract discrete clusters of negotiators' states based on their emotional reactions and spoken words, and can also predict transitions between these states with a certain degree of accuracy. This approach makes a significant contribution to the development of methods for analyzing interpersonal interactions and can be used to optimize processes across various industries, including the banking sector. For example, the developed tool could be employed for automated screening of client states in bank branches, thereby enabling the detection of suspicious changes that may indicate the potential influence of fraudulent schemes.

Despite the successes achieved, the study has several limitations: issues with the synchronization of recordings, a limited sample size, and the specificity of the negotiation scenarios, all of which may affect the reproducibility of the results. These limitations should be considered when interpreting the obtained data and planning future research. In the future, it is recommended to expand the sample, refine data processing methodologies, and test the tool under various conditions to enhance its versatility and practical applicability.

Conclusion

In this study, an evaluation was conducted on the feasibility of applying multimodal embeddings for predicting negotiators' states, their transitions, and the content of negotiations. A multimodal dataset was collected, comprising audio and video recordings of paired negotiations involving 134 participants, which enabled the incorporation of both verbal and nonverbal communication cues. The experimental results indicate that neural network-based models, applied to data clustering using PCA, t-SNE, DBSCAN, and K-means, can extract five discrete clusters representing negotiators' states. These clusters are characterized by varying levels of confidence, uncertainty, willingness to engage, passivity, and emotional expressiveness.

Furthermore, predictive models built using the gradient boosting algorithm demonstrated an accuracy of up to 70% in forecasting state transitions, with metrics such as precision, recall, and ROC AUC indicating a moderate ability to discriminate between positive and negative cases. Feature importance analysis revealed that multimodal embeddings integrating both verbal and emotional information are more significant for predicting state transitions than features derived solely from text.

The results confirm that integrating verbal and nonverbal cues into a unified model enables effective modeling of the dynamic nature of negotiators' states. The proposed approach has practical implications and can be applied for automated analysis of client states across various industries, including the banking sector, to detect suspicious changes potentially associated with fraudulent schemes.

Despite these promising findings, the study has several limitations, including issues with synchronization of recordings, a limited sample size, and the specificity of the negotiation scenarios, which may affect the reproducibility of the results. Future work should focus on expanding the dataset, refining data processing methodologies, and conducting additional tests to enhance the generalizability and practical applicability of the proposed approach.

In summary, the present study demonstrates that multimodal embeddings derived from the analysis of both verbal and nonverbal signals constitute an effective tool for modeling and predicting the dynamic states of negotiators.

References

1. Bern, E. (1961). *Transactional Analysis in Psychotherapy*. New York, NY: Grove Press.
2. Vygotsky, L. S. (1986). *Thought and Language* (Original work published 1934). Cambridge, MA: MIT Press
3. Laclau, E., & Mouffe, C. (1985). *Hegemony and Socialist Strategy: Towards a Radical Democratic Politics*. London: Verso.
4. Foucault, M. (1972). *The Order of Discourse*. New York, NY: Pantheon Books.
5. Darwin, C. (1872). *The Expression of the Emotions in Man and Animals*. London: John Murray.
6. Ekman, P. (1999). Basic Emotions. In T. Dalgleish & M. J. Power (Eds.), *Handbook of Cognition and Emotion*. Chichester, UK: John Wiley & Sons.
7. Plamper, J. (2012). *The History of Emotions*. [Translated version, publisher details as applicable].
8. Russell, J. A. (1980). A Circumplex Model of Affect. *Journal of Personality and Social Psychology*, 39(6), 1161–1178.
9. Altman, I., & Taylor, D. A. (1973). *Social Penetration: The Development of Interpersonal Relationships*. New York, NY: Holt, Rinehart and Winston.
10. Bruner, J. S. (1957). On Perceptual Readiness. *Psychological Review*, 62(2), 123–152.
11. Tiedens, L. Z., & Fragale, A. R. (2003). Power Moves: Complementarity in Submissive and Dominant Nonverbal Behavior. *Journal of Personality and Social Psychology*, 84(3), 558–568.
12. Goffman, E. (1959). *The Presentation of Self in Everyday Life*. New York, NY: Doubleday.
13. Walton, R. E., & McKersie, R. B. (1965). *A Behavioral Theory of Labour Negotiation*. New York, NY: McGraw Hill.
14. Fisher, R., & Ury, W. (1981). *Getting to Yes: Negotiating Agreement Without Giving In*. New York, NY: Penguin Books.
15. Hofstede, G. (2011). Dimensionalizing Cultures: The Hofstede Model in Context. *Online Readings in Psychology and Culture*, 2(1). Retrieved from http://scholarworks.gvsu.edu/orpc/vol2/iss1/8. https://doi.org/10.9707/2307-0919.1014
16. Bazerman, M. H., & Neil, M. (1992). *Rational Negotiations*. Boston, MA: Harvard Business School Press.
17. Khare, A., Parthasarathy, S., & Sundaram, S. (2020). *Self-Supervised Learning with Cross-Modal Transformers for Emotion Recognition*. https://doi.org/10.48550/arxiv.2011.10652
18. Mohamed, Y., Lemaignan, S., Güneysu, A., Jensfelt, P., & Smith, C. (2024). Fusion in Context: A Multimodal Approach to Affective State Recognition. *arXiv.Org*. https://doi.org/10.48550/arxiv.2409.11906

19. Waligora, P., Aslam, M. H., Zeeshan, M., Belharbi, S., Koerich, A. L., Pedersoli, M., Bacon, S., & Granger, É. (2024). *Joint Multimodal Transformer for Emotion Recognition in the Wild.* https://doi.org/10.1109/cvprw63382.2024.00465
20. Praveen, R. G., & Alam, J. (2024). Recursive Joint Cross-Modal Attention for Multimodal Fusion in Dimensional Emotion Recognition. *arXiv.Org.* https://doi.org/10.48550/arxiv.2403.13659
21. Sreevidya, P., Sreevidya, P., & Samiappan, S. (2024). Intermediate Layer Attention Mechanism for Multimodal Fusion in Personality and Affect Computing. *IEEE Access.* https://doi.org/10.1109/access.2024.3442377
22. Kumar, P., Malik, S., & Raman, B. (2022). Hybrid Fusion Based Interpretable Multimodal Emotion Recognition with Insufficient Labelled Data. *arXiv.Org.* https://doi.org/10.48550/arXiv.2208.11450
23. Wu, Y., Zhang, Z., Peng, P., Zhao, Y., & Qin, B. (2022). *Leveraging Multi-modal Interactions among the Intermediate Representations of Deep Transformers for Emotion Recognition.* https://doi.org/10.1145/3551876.3554813
24. Georgescu, A.-L., Chivu, G.-I., & Cucu, H. (2024). *Exploring Fusion Techniques for Multimodal Emotion Recognition.* https://doi.org/10.1109/comm62355.2024.10741464
25. SriHari, M., & Bhaskaran, V. (2024). *A Multimodal Fusion Approach: Emotion Identification from Audio and Video Using Convolutional Neural Network.* https://doi.org/10.1109/cvmi61877.2024.10782687
26. Zhu, Y. (2024). *Advancing Multimodal Emotion Analysis: Integrating Machine Learning and Deep Learning Approaches.* https://doi.org/10.1117/12.3052226
27. Ibrahim, H., Loo, C. K., Geda, S. Y., & Al-Talabani, A. K. (2024). Optimizing Emotional Insight through Unimodal and Multimodal Long Short-term Memory Models. ARO. *The Scientific Journal of Koya University.* https://doi.org/10.14500/aro.11477
28. Wang, X., Ran, F., Hao, Y., Zang, H. L., & Yang, Q. (2024). *Sequence Modeling and Feature Fusion for Multimodal Emotion Recognition.* https://doi.org/10.1109/iccect60629.2024.10546216
29. Filali, H., Boulealam, C., Fazazy, K. E., Mahraz, M. A., Tairi, H., & Riffi, J. (2025). Meaningful Multimodal Emotion Recognition Based on Capsule Graph Transformer Architecture. *Information.* https://doi.org/10.3390/info16010040
30. Bhattacharyya, S., Yang, S., & Wang, J. Z. (2025). *A Heterogeneous Multimodal Graph Learning Framework for Recognizing User Emotions in Social Networks.* https://doi.org/10.48550/arxiv.2501.07746
31. Shi, J., Li, M., Chen, Y., Cui, L., & Bai, L. (2024). Multimodal Graph Learning with Framelet-based Stochastic Configuration Networks for Emotion Recognition in Conversation. *Information Sciences.* https://doi.org/10.1016/j.ins.2024.121393
32. Feng, J., & Fan, X. (2025). *Cross-modal Context Fusion and Adaptive Graph Convolutional Network for Multimodal Conversational Emotion Recognition.* https://doi.org/10.48550/arxiv.2501.15063

33. Zhao, H., Ju, Y., & Gao, Y. (2024). *Bilevel Relational Graph Representation Learning-based Multimodal Emotion Recognition in Conversation.* https://doi.org/10.1109/icme57554.2024.10688301
34. Bhattacharyya, S., Yang, S., & Wang, J. Z. (2025). *A Heterogeneous Multimodal Graph Learning Framework for Recognizing User Emotions in Social Networks.* https://doi.org/10.48550/arxiv.2501.07746
35. Babaev, D., Ovsov, N., Kireev, I., Ivanova, M., Gusev, G., Nazarov, I., & Tuzhilin, A. (2022). CoLES: Contrastive Learning for Event Sequences with Self-Supervision. In *Proceedings of the 2022 International Conference on Management of Data* (pp. 1190–1199). ACM. http://dx.doi.org/10.1145/3514221.3526129
36. Clark, K., Luong, M.-T., Le, Q., & Manning, C. (2020). *ELECTRA: Pre-training Text Encoders as Discriminators Rather Than Generators.* arXiv preprint arXiv:2003.10555
37. Girdhar, R., et al. (2023). *ImageBind: Connecting Text, Image, Audio, and Video.* arXiv preprint arXiv:2305.05665.
38. Zhang, Y., et al. (2023). *Meta-Transformer: A Unified Framework for Multimodal Learning.* arXiv preprint arXiv:2307.10802.
39. Radford, A., Kim, J. W., Xu, T., Brockman, G., McLeavey, C., & Sutskever, I. (2022). *Robust Speech Recognition via Large-Scale Weak Supervision.* arXiv preprint arXiv:2212.04356.

Chapter 6

Simulating Dynamics of the Recognition-Primed Decision Model on Financial Markets Using LLM

Yana Ivanova, Alexander Didenko,
Mikhail Grigoryan, and Artem Kraevskiy

Introduction

The financial sector offers a compelling case study for both the evolution of artificial intelligence (AI) technologies and the displacement or augmentation of the human workforce with AI. The evolution of algorithmic and AI technologies in financial markets has unfolded through a series of distinct epochs, each characterized by incremental advancements in computational methods and trading systems. In the early phase, spanning the 1980s to the 1990s, financial institutions began transitioning from manual trading floors to electronic platforms by implementing rule-based systems and expert systems. During this period, pre-programmed rules were employed for trade execution and market analysis. For example, the New York Stock Exchange introduced the Designated Order Turnaround (DOT) system in 1976 and later enhanced this capability with the SuperDOT upgrade in 1984, thereby increasing order size and execution speed. Concurrently, regulatory changes

 DOI: 10.4324/9781003606956-6

in London, particularly the "Big Bang" deregulation of 1986, prompted the shift from open-outcry to screen-based trading. This development accelerated the global adoption of electronic trading platforms and set the stage for the later emergence of sophisticated algorithmic systems. The events of Black Monday in 1987 further underscored both the promise and the perils of automated systems, as program trading based on fixed rules contributed to a cascade of sell orders during a market crash. In addition, early statistical arbitrage strategies that exploited pricing inefficiencies, such as pairs trading, demonstrated the potential of algorithmic models to outperform human judgment, thereby inspiring the creation of quantitative hedge funds. The proliferation of online trading platforms in the early 1990s further diminished reliance on human intermediaries and laid the foundation for increasingly complex automated systems.

The subsequent epoch, spanning the 1990s to the 2000s, witnessed a paradigm shift as machine learning (ML) techniques and advanced statistical models began to replace static rule-based systems with dynamic models capable of learning from data [1]. The founding of The Prediction Company in 1991 marked an early instance of applying ML to forecast market movements and adapt trading strategies. Neural networks were employed to recognize complex patterns in historical data, and ensemble methods such as boosting and bagging, as well as genetic algorithms and simulated annealing, were integrated to optimize portfolio performance and assess credit risks [2]. At the same time, high-frequency trading (HFT) emerged as a significant force in financial markets. Enabled by advancements in computational power and electronic exchanges, HFT strategies executed trades in fractions of a second [3].

In the 2010s, the integration of big data and alternative data sources fundamentally transformed financial analysis and trading strategies. Financial institutions increasingly harnessed unconventional data streams—such as satellite imagery, social media feeds, and web-scraped information—to generate trading signals and improve market forecasts. The advent of natural language processing (NLP) techniques enabled the automated analysis of news articles and earnings call transcripts, enhancing the accuracy and speed of sentiment analysis [2]. During this period, the launch of robo-advisors brought automated portfolio management to retail investors, while institutional players integrated vast arrays of alternative data into risk management frameworks and strategic research initiatives. The development of in-house data laboratories and analytics platforms, exemplified by systems like BlackRock's Aladdin, underscored the increasing reliance on data-driven decision-making in finance.

From the late 2010s into the early 2020s, the deployment of deep learning, reinforcement learning, and graph-based analytics further enhanced the sophistication of AI systems in financial markets. Deep learning models were applied to capture nonlinear relationships within market data and to improve predictive accuracy in complex tasks such as option pricing. Reinforcement learning allowed trading algorithms to iteratively learn optimal strategies through trial and error in real-time market environments [4]. Simultaneously, graph neural networks were introduced to analyze interconnected data—such as relationships among market participants—and to detect fraudulent activities. This period also witnessed the emergence of autonomous hedge funds, with examples like Aidyia illustrating the potential of systems that combine deep learning and evolutionary computing to execute trades without human intervention.

The current era, defined by the rise of large language models (LLMs), represents a significant leap in the integration of AI into financial systems. AI is rendered at all levels of autonomy as either independent trading or execution bots, decision-support assistants, or collaborative co-pilots. Institutions have begun to deploy proprietary LLMs, such as BloombergGPT and adaptations of GPT-4, to process and interpret vast quantities of textual data. Tasks considered routine, such as order execution and compliance-related reporting, are progressively executed by AI bots that function with limited human intervention. On the other hand, AI assistants play a pivotal role in decision-making support by formulating recommendations, as illustrated by the internal deployment of GPT-4 applications at Morgan Stanley. The nascent co-pilot model epitomizes a synergistic engagement where AI and human operators partake in an ongoing iterative process—each party contributing to and enhancing the ultimate decision [5]. These models enhance real-time analysis and support decision-making across a range of functions including trading, strategic market research, and regulatory compliance. Financial firms are leveraging LLM-based applications to transcribe and analyze client communications, extract actionable insights, and facilitate the rapid generation of trade ideas [6]. The transition from simple rule-based systems to autonomous, AI-driven architectures encapsulates the broader trend of integrating advanced computational intelligence into finance—a progression that is directly relevant to the development of derivatives algorithmic trading systems based on LLMs, which is a focus for the current chapter.

Derivatives Trading

In the context of financial markets, a derivative is a contractual instrument whose value is derived from the performance of an underlying asset, index, rate, or event. These underlying entities can include stocks, bonds, commodities, currencies, interest rates, market indices, or even economic indicators. Derivatives are primarily utilized by institutional investors, such as hedge funds, for purposes of hedging risk, speculating on price movements, or generating income through premium collection, without necessarily owning the underlying asset. Derivatives are categorized into several types, including options, futures, forwards, and swaps. Vanilla (or simple) options grant the holder the right, but not the obligation, to buy (call option) or sell (put option) the underlying asset at a specified price (strike price) before or at expiration. Options could be used in combinations—or "strategies"—to gain more control over risk exposure, and to provide flexibility to the trader in implementing his market view with a portfolio of financial instruments.

Option is characterized by a strike price, which is the predetermined price at which the holder can buy or sell the underlying asset, and expiration date, which determines time constraint on the right embodied in the option contract. The cost of acquiring this right is the premium, which consists of intrinsic value—determined by the option's moneyness—and time value, reflecting market expectations of future volatility. The relationship between the strike price of an option and the current market price of the underlying asset is described by the term "moneyness". An option can be in one of three states: in the money (ITM) if exercising it would result in an immediate profit. A call option is ITM when the market price of the underlying asset is above the strike price. A put option is ITM when the market price is below the strike price. An option is at the money (ATM) if the market price of the underlying asset is equal to the strike price. In this case, the option has no intrinsic value but still retains time value. An option is out of the money (OTM) if exercising it would result in a loss. A call option is OTM when the market price is below the strike price, while a put option is OTM when the market price is above the strike price. Sensitivity of the options to various factors is measured by so-called "greeks". Delta measures the rate of change in the option's price relative to the underlying asset's price. Gamma indicates the rate of change of Delta, capturing the non-linearity of price movements. Theta represents time decay, showing how an option loses value as expiration approaches. Vega measures sensitivity to changes in implied volatility, affecting the option's price even if the underlying asset remains static. Rho quantifies the impact of interest rate changes on the option's valuation. Greeks, moneyness, type of the contract (put/call), and side of the trade (buy/sell) are key factors in constructing complex strategies.

To illustrate a complex strategy, consider a case where a trader believes that the SP500 stock market index will remain within a relatively narrow range over the next month. In such a scenario, the trader may opt for an iron condor strategy to articulate a nuanced view of low volatility. The iron condor is constructed by selling an out-of-the-money (cheap) call and an out-of-the-money put while simultaneously purchasing further (in terms of expiration date) out-of-the-money options on both sides. This combination allows the trader to collect premiums from the sold options, thereby benefiting from time decay as long as the SP500 index remains confined within the selected strike prices. When the trader is limited to using only spot trades or simple vanilla derivatives, the opportunity to capture premium income and effectively manage risk under conditions of limited price movement is significantly reduced. Alternatively, if the trader anticipates a powerful directional move in the SP500—whether upward or downward—a strangle strategy becomes more appropriate. The strangle involves the purchase of an out-of-the-money call and an out-of-the-money put, a structure that requires an initial premium outlay but offers the potential for considerable profit if the market moves decisively. This approach is particularly effective when large swings in the index are expected, as it provides an asymmetric payoff profile; the gains on one side can more than offset the total premium paid. In contrast, simple spot trades or basic vanilla options do not afford the same opportunity to capitalize on significant market volatility.

Naturalistic Decision-Making and Recognition-Primed Decision Model

Naturalistic Decision-Making (NDM) is a framework for understanding decision-making in real-world settings characterized by complexity, uncertainty, incomplete information, and time constraints. Unlike classical decision-making models that rely on exhaustive evaluation of alternatives, NDM posits that experts use experience, heuristics, and pattern recognition to make decisions efficiently [7]. The emergence of NDM as a distinct paradigm in decision science stems from studies of experts operating under high-stakes conditions, where rapid decision-making is required despite limited data. One of the central models within NDM is the Recognition-Primed Decision

(RPD) model, developed by Gary Klein. The RPD model suggests that experienced decision-makers rely on pattern recognition to identify a plausible course of action without systematically comparing multiple options [8].

The concept of bounded rationality, introduced by Herbert Simon, complements NDM by arguing that human decision-makers operate within cognitive and informational constraints that limit their ability to optimize [9]. NDM emphasizes the role of experience and intuition, where decision-makers rely on patterns and tacit knowledge acquired through experience to make effective decisions quickly. In financial markets, this is particularly relevant, as professionals operate in ill-structured and indeterminate environments, requiring them to act as "satisficers" rather than optimizers—aiming for acceptable rather than optimal choices.

Behavioral economics further supports this perspective by demonstrating how biases and heuristics, such as loss aversion and the herd effect, influence market behavior, leading to anomalies like the momentum effect and long-term reversals. Decision-making in financial markets has traditionally been examined through the lens of classical economic theories, which assume rational actors optimizing utility in a well-structured environment. However, real-world financial decisions are made under uncertainty, time pressure, and incomplete information, necessitating a framework that accounts for human cognitive limitations and expertise. NDM provides such a framework, explaining how financial professionals—including traders and portfolio managers—make decisions in complex and dynamic settings. Rather than calculating all possible outcomes, financial professionals often employ "satisficing" strategies, selecting the first available option that meets a predefined threshold of acceptability [10].

This approach aligns with empirical observations of traders, who frequently make rapid decisions based on partial information rather than conducting exhaustive analyses. Rather than engaging in deliberate analytical processing, these professionals frequently depend on prior experiences to recognize familiar situations and execute responses accordingly. Intuition also plays a critical role in NDM. Kahneman and Klein [11] distinguish between expert intuition and heuristic-driven biases, noting that in structured environments with recurrent patterns, experts develop reliable intuitions based on accumulated experience. In financial contexts, this translates to traders and fund managers leveraging their tacit knowledge to interpret market signals, anticipate price movements, and assess risk.

Empirical research has substantiated the presence of NDM mechanisms in financial decision-making. A study by Coates found that traders with higher interoceptive awareness—measured by their ability to sense their own heartbeats—exhibited superior trading performance and longevity in the industry. This finding suggests that physiological intuition plays a role in risk assessment and decision execution under high-stress conditions.

Observational studies of trading desks provide further evidence of NDM in practice. Foreign exchange traders, for example, develop heuristic-based decision strategies through exposure to repetitive market patterns. Rather than performing exhaustive calculations, traders often rely on mental shortcuts, such as anchoring to historical price levels or reacting to real-time market sentiment. Ethnographic research has also highlighted the use of analogical reasoning, where investors draw on past experiences to make judgments about new situations. This aligns with the RPD framework, in which experts recognize situational cues that map onto previously encountered scenarios [12].

Computational RPD Modeling

There is a longstanding tradition of modeling RPD processes in digital environments. Nowroozi et al. extended the original Klein model by incorporating cognitive architectures, such as Soar, to further simulate human-like decision-making [13]. Ji et al. [28] introduced a fuzzy logic-based computational RPD model that enhances decision-making under uncertainty by formalizing subjective information through fuzzy sets and similarity measures. This model was tested in medical decision-making, particularly in assessing drug adverse effects, demonstrating improved agreement with human experts. Hutton et al. explored RPD applications in computer-generated forces (CGF), modeling adversary decision-making in military simulations. Their implementation emphasized satisfaction over optimization, improving realism in tactical training environments. Hu et al. further refined RPD applications by integrating Bayesian networks to predict pilot behavior in midair encounters, highlighting the effectiveness of probabilistic approaches in decision-making under stress. Norling et al. proposed integrating RPD into belief-desire-intention (BDI) architectures to enhance realism in multi-agent simulations [13]. The BDI framework is widely used in multi-agent systems to simulate human cognitive processes. They demonstrated that agents using RPD-based heuristics could better approximate human decision-making compared to purely rational models. Fan et al. extended this approach by incorporating human-agent collaboration, creating an adaptive decision-making system (R-CAST) that improves team performance under time pressure [14]. This work showcased how expectancy monitoring and information-sharing strategies could optimize decision outcomes in dynamic environments. Hassard examined the role of RPD variations in design decision-making, identifying three distinct approaches: simple recognition, refining ambiguous situations, and dynamically adjusting solutions. His findings emphasized that mental simulation plays a critical role in evaluating design alternatives, aligning with findings in cognitive psychology.

Despite significant advancements, several challenges remain. Ji et al. highlighted the difficulty of encoding expert knowledge into computational RPD models, particularly in medical applications. Hassard noted that existing RPD frameworks struggle to account for persistent analogies in design decision-making [15]. Hu et al. emphasized the need for improved uncertainty management techniques in RPD-Bayesian network models. Hutton et al. suggested incorporating cognitive task analysis to refine expectancies in RPD-based simulations. These advancements in computational RPD modeling underscore its applicability across diverse domains, from military simulations to medical decision-making.

Building on this foundation, the next section explores how the principles of RPD and NDM can be extended to autonomous trading systems, leveraging large language models to enhance decision-making in financial derivatives markets.

The intersection of NDM, RPD models, and financial derivatives trading provides a compelling framework for the application of large language models (LLMs) in autonomous decision systems. The complexity of derivative-based strategies necessitates rapid, experience-driven judgments, often made under uncertainty and time constraints—conditions that align closely with the principles of NDM and RPD. By leveraging LLMs within this paradigm, it becomes possible to construct systems capable of dynamically managing portfolios, identifying high-probability market patterns, and executing structured yet flexible trading strategies. Given the inherent difficulties in encoding tacit knowledge and heuristic-driven decision-making within traditional algorithmic approaches, the integration of LLMs offers a novel means of simulating expert intuition. This confluence of financial markets, NDM theory, and LLM-driven automation serves as the foundation for the methodology detailed in the following section, where we introduce both the NDM-Fin v.1.0 benchmark and the autonomous system prototype designed to execute RPD-style reasoning in the context of derivatives trading.

Method

The backtest procedure was organized as follows. First, we've taken several databases, covering financial news and macroeconomic data, and using multi-step summarization with LLM compressed it into time-stamped semantically coherent units, storing content that is relevant for option trading tasks, thus making NDM-Fin an evaluation benchmark. This static database, as well as database of S&P 500 option chains data, was used in all further simulations [16]. Then, we run a simulation backtest with daily granularity, feeding market outlooks, options chains data, and updated portfolio states to the trading bot for recognition of opportunities for portfolio rebalancing or opening new positions. We describe this process in more detail further.

NDM-Fin v.1.0 Preparation

Making up the LLM bot often starts with making preparation of evaluation benchmarks. We begin construction of our benchmark with the ingestion and preprocessing of raw textual financial data, consisting of news articles, to remove noise and standardize formats. Next, the data is segmented into smaller, semantically coherent chunks, which are then embedded using a multilingual embedding model that captures contextual relationships. Parallel to this, a large language model is used to identify and generate keywords or short phrases indicative of early warning signals—these terms focus on events such as unexpected price moves, sudden rate adjustments, abrupt volatility swings, and other phenomena likely to trigger notable shifts in market sentiment. Once all text chunks and keywords have been converted to their respective embeddings, they are stored in a vector database [17], allowing semantic similarity queries. At a daily resolution over rolling seven-day windows, relevant chunks are fetched from the database by matching their embeddings to the early warning signal terms. These chunks are then passed through a multi-step summarization procedure, using a chain-of-thought approach with map-reduce-style aggregation, resulting in a concise composite textual overview of key macroeconomic events, sector-level risk events, and volatility signals. The full pipeline is presented in Figure 6.1.

We classify the general market direction based on the position of the S&P 500 index relative to its 200-day simple moving average (SMA). If the current S&P 500 price is above the 200-day SMA, we categorize the market outlook as upward. Conversely, if the current price is below the 200-day SMA, we categorize it as downward. This approach aligns with conventional

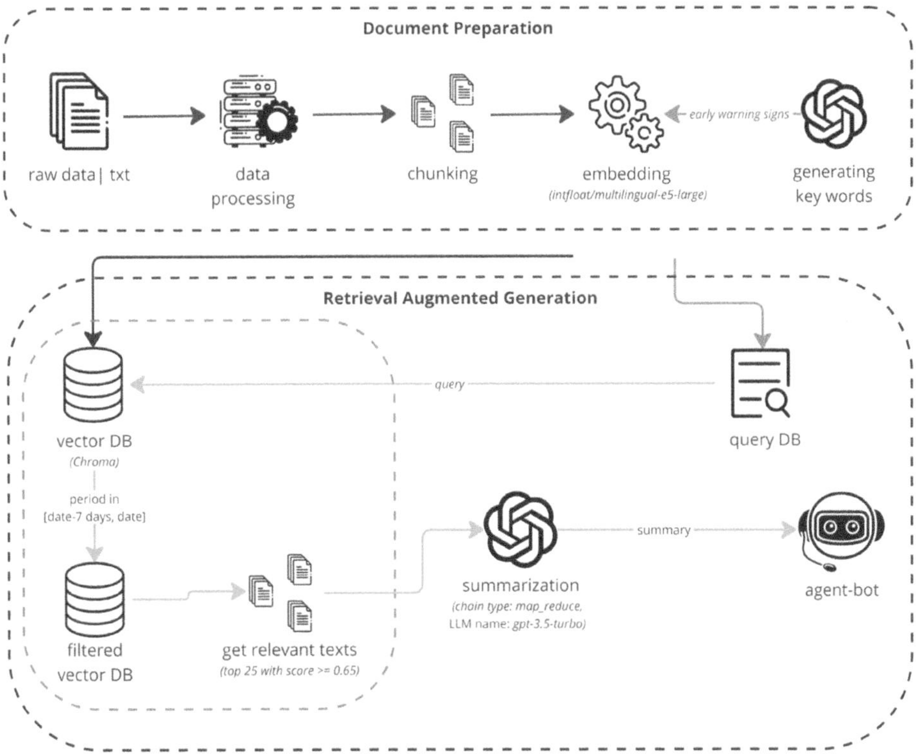

Figure 6.1 Pipeline for NDM-Fin 1.0 preparation.

trend-following methodologies used in technical analysis, where the 200-day SMA serves as a widely recognized threshold for distinguishing between bullish and bearish market conditions. Volatility classification is determined using the CBOE Volatility Index (VIX), which reflects market expectations for near-term volatility [18]. We define three distinct volatility regimes based on fixed threshold levels of VIX:

- Low volatility: VIX is below 15, indicating a stable market with limited price fluctuations.
- Medium volatility: VIX ranges between 15 and 25, representing a moderate level of market uncertainty.
- High volatility: VIX exceeds 25, signaling heightened uncertainty and increased market risk.

By categorizing market conditions along the dimensions of trend direction and volatility, we can assess how well the system adapts its decision-making across different market regimes. This enables a structured evaluation of its ability to recognize patterns, adjust strategy selection, and optimize portfolio management under varying levels of uncertainty.

Main Trading Cycle

The daily cycle begins by aggregating new information from three main sources—news summaries, macroeconomic data summaries, and option prices—to capture the evolving market landscape. For simulation we use a pre-generated static market outlook evaluation benchmark, described earlier. Market outlook inputs are then fed into an initial "perception" prompt (LLM #1), which analyzes the raw data and produces a concise representation of relevant decision-making cues and market expectancies in JSON format. This structured output is subsequently passed to a second large language model (LLM #2) with a "course-of-actions" prompt, whose role is to identify an appropriate candidate trading strategy from a preexisting library of structured strategy templates [19]. This contrasts with the canonical Recognition-Primed Decision-Making (RPDM) formula, according to which "recognition has four by-products": cues, plausible goals, course-of-action plan, and expectancies. In our case, we substitute one-shot simultaneous recognition with four by-products with two-shots process, having two by-products (cues and expectancies) after the first shot, and course-of-action with steps pre-baked with plausible goals, after the second.

The course-of-action model selects one option strategy and justifies its choice based on the prevailing macro signals, news themes, and observed option price movements. Next, the chosen strategy is passed to a third, "critic and execution", large language model (LLM #3) that evaluates the available option contracts—only those within a preset range of strikes and expiration dates—and proposes specific trades following a standardized template. These candidate trades are then subjected to a "logocentric" decision step, which determines whether the level of conviction warrants execution. If deemed sufficient, the trade is dispatched to an automated execution bot, otherwise the system concludes its cycle. This approach imitates the market impact assessment that institutional investors typically undertake before finalizing a trade [20]. In practice, large-scale market participants often conduct pre-trade analytics to gauge potential price slippage, liquidity constraints, and broader market ripple effects, ensuring that the projected risk–reward profile remains favorable. By incorporating a "logocentric" decision step at the end of the pipeline, our system similarly evaluates whether anticipated transaction costs and volatility align with internal thresholds. If these criteria are met, the final trade instructions are handed off to the execution bot; otherwise, the system withholds the order until further data or a more opportune market environment emerges. This iterative process thus approximates the institutional mindset, balancing strategy selection with real-world constraints around market impact and trade feasibility.

By iterating this pipeline daily (as shown in Figure 6.2), the framework integrates updated market data with systematic strategy selection and risk management, offering a coherent, end-to-end decision process for automated option trading.

Results

Figure 6.3 presents the performance of the basic LLM-RPD trading agent under different benchmark conditions, categorized by market volatility (low, medium, and high) and price direction (upward or downward). The boxplots illustrate the distribution of profit and loss (PnL) for each scenario, revealing key insights into the model's behavior across different market regimes.

In low-volatility environments, the model performs consistently well when the market trends upward, yielding positive PnL values with minimal

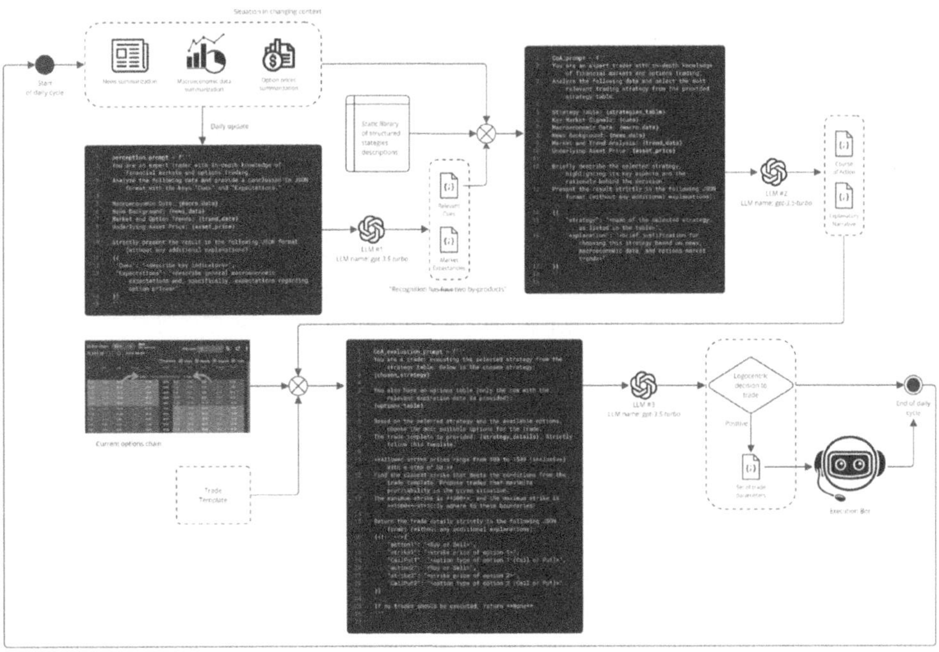

Figure 6.2 Pipeline for daily trading cycle of the bot evaluation.

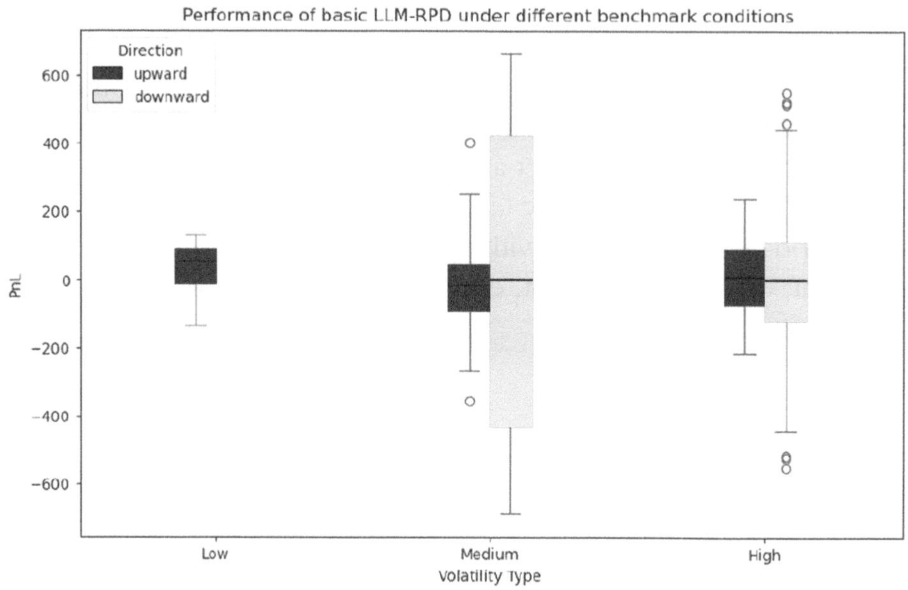

Figure 6.3 Performance of basic LLM-RPD system on NDM-Fin v.1.0.

dispersion. However, the downward market case for low volatility is absent, suggesting that the bot might not have engaged in trades or that the dataset lacked sufficient downward moves in stable conditions.

For medium volatility, the model exhibits significantly greater variance, especially in downward market conditions. The distribution of PnL in the downward case spans a wide range, indicating high sensitivity to market fluctuations. The median return for the downward scenario is lower than in upward conditions, confirming that the bot struggles to maintain stability in such conditions. The numerical results support this, as the mean PnL for downward trades in medium volatility is −13.89, whereas in upward trends, it is slightly higher at −13.21, but still negative, reflecting overall performance deterioration in increased volatility environments.

In high-volatility scenarios, both upward and downward markets exhibit large return dispersion, with visible outliers. The model shows positive median performance in both cases, but the overall variance suggests that while it can capitalize on extreme moves, it also faces substantial downside risks [21]. The numerical breakdown confirms that in high-volatility upward markets, the model achieves a 9.98 PnL, whereas in downward cases, it yields a 2.96 PnL, suggesting that its ability to handle volatile downward trends remains weak relative to upward markets.

Overall, the results indicate that the LLM-RPD model performs best in stable upward-trending markets but struggles in medium-volatility downward trends, where PnL is significantly negative. In high-volatility conditions, the model exhibits a wider range of performance, suggesting that its adaptability to extreme market moves requires further refinement.

Discussion

The evolution of artificial intelligence and robotics over the past several decades has produced a rich landscape of automated systems—from early rule-based trading bots to contemporary autonomous agents leveraging large language models (LLMs) for logocentric decision-making, and recent advances in domain-specific LLMs like BloombergGPT and FinGPT. This chapter makes a contribution to the field by introducing a novel bot architecture that integrates LLM-based reasoning to achieve sophisticated, language-driven decision processes. We choose the financial markets as the domain for illustrating our LLM-driven decision-maker due to their relatively well-defined structure, which simplifies the construction of an initial proof-of-concept. In such settings, an agent's actions are confined to a limited repertoire—primarily issuing buy and sell orders—while its perception of both internal states (e.g., portfolio composition and current profit/loss profile) and external signals (macro indicators, technical metrics, unstructured textual data, and overall market sentiment) remains constrained yet sufficiently rich to test sophisticated decision-making. Despite this structural clarity, financial markets exhibit abrupt shifts and high volatility, offering an ideal environment in which to evaluate the robustness and adaptability of language-driven agents [22]. The controlled yet dynamic nature of the domain allows for the practical exploration and validation of various LLM-centric decision-making architectures. Moreover, the industry provides well-established performance metrics, such as the Sharpe ratio and drawdowns, and readily available baselines—e.g., major market indices—against which to benchmark. These standardized measures of success and failure make finance an especially suitable testbed for iterative development and evaluation of LLM-driven reasoning engines.

The choice of the Recognition-Primed Decision (RPD) model as the core decision-making framework for our LLM-driven trading agent is motivated by the inherent characteristics of financial markets, which demand rapid, experience-driven reasoning under uncertainty. Unlike classical models of rational decision-making that assume exhaustive evaluation of alternatives, RPD aligns with the way expert traders operate in real-world conditions [10, 19]. In derivatives trading market participants rarely render rigorous first-principles analysis; instead, they rely on pattern recognition, heuristics, and situational awareness to make high-stakes decisions in real time. RPD, developed by Gary Klein, provides a structured model of this intuitive, experience-based decision process, making it particularly suitable for the development of autonomous trading agents that must respond dynamically to evolving market conditions [23].

We specifically focus on complex options strategies rather than simpler financial instruments due to their suitability for testing the limits of LLM-driven decision-making in a structured yet nontrivial environment. Unlike spot trading, where decision-making is often reduced to binary choices—buy, sell, or hold—derivative trading requires a deeper level of strategic reasoning. Options strategies involve nonlinear payoffs, interactions between multiple contracts, and sensitivity to factors beyond mere price direction, such as implied volatility and time decay. This complexity makes them an ideal domain for evaluating the reasoning capabilities of LLMs, as successful strategy selection necessitates multi-step planning, scenario analysis, and probabilistic reasoning. Additionally, options markets are heavily influenced by sentiment and macroeconomic events, requiring the agent to integrate both structured numerical data and unstructured textual information into its decision-making process.

From a practical standpoint, institutional market participants require not just an AI-driven trading system but an interpretable AI—a system that not only executes decisions but also provides transparent reasoning behind its actions [24]. In financial markets, where capital allocation decisions involve substantial risk, black-box models lacking explainability are often unsuitable for institutional adoption due to regulatory, fiduciary, and operational constraints. Interpretability ensures that traders, risk managers, and compliance officers can audit the model's rationale, assess its consistency with established investment frameworks, and trust its decision-making under varying market conditions.

The Recognition-Primed Decision (RPD) model is particularly well-suited for this requirement, as it provides a structured decision-making process that is inherently interpretable. Unlike purely statistical machine learning models that generate predictions without offering insight into their underlying logic, an RPD-based system explicitly maps decisions to recognizable patterns and previously encountered scenarios, mimicking the cognitive workflow of expert traders. By structuring an AI system around RPD principles, we enable it to explicitly articulate reasoning in the most human-friendly way possible—by producing a narrative, or compelling story, which is easy to understand, and communicate further, and which allows for more flexibility than traditional approaches like, say, SHAP values [25]. In this proof-of-concept we also show how persistent analogies in decision-making can be mitigated through dynamic RAG- and chain-of-thought enabled context-aware prompt engineering, allowing LLMs to dynamically refine decision heuristics rather than relying on static analogy sets. Wrapping Klein's RPD elements (action plans, cues, etc.) in JSON format illustrates the value of integrating structured communication protocols (e.g., hybrid natural language and templated reports) to mitigate the "telephone effect" observed in purely dialogue-based systems [7].

Ultimately, we present this study as an illustration of a broader approach—namely, a logocentric Recognition-Primed Decision (RPD) model system capable of executing actions within a defined domain [26]. While this chapter focuses on financial markets as a controlled yet dynamic setting, where actions take place entirely in the digital domain, we believe that the integration of LLM-based reasoning with RPD is applicable beyond algorithmic trading. The underlying mechanism—leveraging natural language as an interface for structured yet flexible decision-making—can extend to physical domains, such as robotics, where autonomous agents must make rapid, experience-driven decisions in real-world environments.

In this broader perspective, LLM-powered RPD systems could serve as cognitive cores for universal home assistants, industrial automation systems, or autonomous vehicles, where decisions require a balance of learned heuristics, situational adaptation, and explainability. Just as our trading agent evaluates structured financial data and unstructured textual information to construct a human-interpretable decision narrative, a robotic system could integrate multimodal sensory input—visual, auditory, and textual data—into a coherent, adaptable framework for decision-making. By grounding AI-driven reasoning in the structured intuition of RPD, we envision a new class of autonomous systems capable of operating in both digital and physical environments, producing interpretable action rationales while dynamically adapting to complex, uncertain conditions [27].

Several promising research directions emerge from this work, extending beyond the immediate application of LLM-driven RPD systems in financial markets. First, a more granular evaluation of decision-making quality remains an open challenge. While traditional financial performance metrics such as the Sharpe ratio or maximum drawdown provide useful benchmarks, they do not directly assess the quality of reasoning itself. Developing new interpretability and benchmarking tools tailored to LLM-driven decision-making systems, possibly by quantifying coherence, consistency, and adaptability in generated decision rationales [28], represents a critical next step. Further improvement of bot prompting architecture might include ReAct-like prompting to ensure explainable, audit-friendly decision trajectories.

A second promising direction for future research is the integration of reasoning-based large language models (LLMs) within reinforcement learning (RL) frameworks, particularly in the context of financial markets. Recent advancements, such as DeepSeek-R1 [4], have demonstrated that LLMs can be effectively trained through RL to develop emergent reasoning behaviors without reliance on traditional supervised fine-tuning. The success of such models underscores the potential for applying RL methodologies to environments where decision-making occurs under uncertainty, and where feedback is directly available from the system dynamics [29]. Financial markets provide an ideal testbed for RL-based reasoning models, as they naturally generate a continuous stream of reward signals through portfolio performance metrics. Unlike domains where human-annotated rewards are required for training, financial markets offer self-contained feedback loops in the form of profit and loss, Sharpe ratios, and drawdowns. This aligns well with recent RL-based improvements in LLM reasoning, such as the S2R framework [13] or SWE-RL, where models were trained to recover reasoning processes from evolving open-source software data. A similar methodology can be applied to trading strategies, where the LLM learns optimal decision paths by iteratively adjusting its approach based on market conditions, leveraging reinforcement learning with market-aware reward models. This is particularly relevant in financial contexts, where decision errors can be identified through adverse portfolio movements and subsequently adjusted through policy optimization techniques. Further research could explore self-correcting financial agents in the style of RL-enhanced LLMs that continuously refine their decision logic by analyzing historical market behavior, economic indicators, and execution outcomes, much like human traders adapt their strategies based on prior experience.

Also, an important research direction involves exploring multi-agent architectures, where multiple specialized large language models (LLMs) collaborate within a single decision-making system. LLM-driven RPD systems could move beyond monolithic prompting chains and instead benefit from distinct, interacting cognitive modules. A relevant example is the TradingAgents framework proposed by Xiao [22], which employs separate agents for market perception, strategy selection, and risk management. Structured coordination models like TradingAgents or MetaGPT [7] align with Marvin Minsky's seminal theory in *The Society of Mind* [16], which posits that intelligence is not a singular entity but rather an emergent property of simpler interacting agents, each responsible for a specific cognitive function. By distributing tasks across specialized agents, such architectures replicate the way biological and artificial intelligent systems achieve complex reasoning through modular expertise and inter-agent collaboration. A trading bot leveraging multi-agent LLM architectures could function as a co-pilot, assisting human traders by dynamically adapting to market conditions, providing real-time strategic recommendations, and mitigating risks. This hybrid human–AI collaboration represents a promising research avenue, potentially leading to more effective decision-making systems in high-stakes financial environments [30].

Recent advancements in LLM-driven decision-making models offer promising solutions to longstanding challenges identified in earlier computational RPD research. The difficulty of encoding expert knowledge into decision-making models can be addressed through self retrieval-augmented generation (RAG), enabling LLMs to extract and apply tacit knowledge from vast financial, medical, or other datasets. Our study made the first attempt of this kind, leveraging local knowledge base on complex derivatives trading strategies, but definitely there is plenty of room for further development, given recent advances in complex RAG architectures. Uncertainty management can be significantly improved by integrating probabilistic inference techniques, such as Bayesian decision models, within LLM-driven architectures, thereby enabling more adaptive responses to evolving environments. Also, fusing fuzzy-like symbolic reasoning with LLM-driven RPD systems could both bring innovative results and show improvement in performance. By leveraging these advancements, LLM-driven RPD models bridge the gap between heuristic-driven human intuition and scalable, data-driven automation, paving the way for more robust applications in autonomous trading, real-time military strategy, and high-stakes medical diagnostics.

Another important research front concerns real-time adaptations of logocentric AI systems. While our approach leverages pre-trained LLMs, integrating dynamic self-improvement techniques can significantly boost system plasticity. For instance, employing episodic memory mechanisms [18] enables the model to continually refine its reasoning pathways. This dynamic self-improvement might be complemented by advances in embodied AI and real-world interaction, where robust action planning [2, 5] facilitates decision-making under conditions of partial observability and rapid contextual shifts. In such contexts, RPD-based reasoning could guide physical interactions with the environment, allowing robots or autonomous systems to refine their decision-making over time based on real-world experience.

A final key area for future exploration is the fusion of multimodal data sources within an LLM-driven Recognition-Primed Decision-Making (RPDM) framework. While this paper has focused on integrating structured numerical indicators and textual financial news to support market forecasting applications, extending this integration to include real-time sensor data, video feeds, and geospatial information presents significant opportunities for broader applications. For instance, incorporating high-dimensional sensor readings and streaming video data could enhance autonomous navigation systems by enabling them to dynamically calibrate their decision-making processes in response to continuously changing physical conditions. Similarly, integrating geospatial data into such a framework could advance smart infrastructure management, leading to more efficient urban planning and improved emergency response systems. Additionally, security applications could benefit from real-time situational awareness provided by multimodal fusion, improving threat detection and response capabilities.

This multimodal integration within the RPDM paradigm is particularly compelling because LLMs excel at synthesizing heterogeneous data sources while simultaneously generating human-interpretable rationales that explain the underlying decision-making process. This capability is essential in domains such as autonomous navigation and security, where transparency, trust, and accountability are critical. Moreover, recent advancements in retrieval-augmented generation, as discussed in [5], highlight the importance of efficiently retrieving diverse knowledge sources and strategically integrating multi-step reasoning to optimize multimodal tasks.

By fusing dynamic sensor inputs, visual signals, and spatial analytics, future RPDM-based systems could significantly improve adaptability and robustness in environments characterized by high uncertainty. This extended integration would reinforce the shift from static, pre-trained AI systems to fully adaptive models—capable of real-time learning and targeted performance adjustments across diverse operational domains. While financial markets serve as a well-defined testbed for demonstrating the feasibility of RPDM-based decision-making, the broader objective of this research is to establish a generalizable LLM-RPDM framework applicable to both digital and physical domains. By leveraging multimodal fusion, we move closer to creating truly autonomous agents capable of complex, real-time reasoning and action across a wide range of dynamic environments.

Taken together, these research directions highlight the broader significance of this work beyond financial markets. By combining LLMs with structured cognitive models such as RPD, we move toward AI systems that are not only autonomous but also interpretable, adaptive, and capable of decision-making in both digital and physical environments.

References

1. Benaroch, M., & Dhar, V. (1995). Controlling the complexity of investment decisions using qualitative reasoning techniques. *Decision Support Systems, 13*(2), 109–128. https://doi.org/10.1016/0167-9236(94)00031-M
2. Chen, Y., Liu, J., & Wang, Y. (2025). Reinforcement learning for long-horizon interactive LLM agents. *arXiv.* https://arxiv.org/abs/2502.06975
3. Cohen, M. S., Freeman, J., & Thompson, B. B. (2017). Training the naturalistic decision maker. In C. E. Zsambok & G. Klein (Eds.), *Naturalistic Decision Making* (pp. 257–278). CRC Press. https://doi.org/10.4324/9781315095080-11
4. DeepSeek-AI, et al. (2025, January 22). DeepSeek-R1: Incentivizing reasoning capability in LLMs via reinforcement learning. *arXiv.* https://arxiv.org/abs/2501.01457
5. Gonzalez-Pumariega, A., Smith, L., & Johnson, M. (2024, December 9). Query-efficient planning with language models. *arXiv.* https://arxiv.org/abs/2501.01457
6. Gore, J., Ward, P., Conway, G. E., Ormerod, T. C., Wong, B. L. W., & Stanton, N. A. (2018). Naturalistic decision making: Navigating uncertainty in complex sociotechnical work. *Cognition, Technology & Work, 20*(3), 575–589. https://doi.org/10.1007/s10111-018-0531-x
7. Hong, Y., Chen, M., & Zhang, L. (2023). MetaGPT: Structured communication protocols for multi-agent systems. *arXiv.* https://arxiv.org/abs/2308.00352
8. Kaczynski, D., Salmona, M., & Smith, T. J. (2014). Qualitative research in finance. *Australian Journal of Management, 39*(2), 127–146. https://doi.org/10.1177/0312896212469611
9. Kesh, S., & Raja, M. (2005). Development of a qualitative reasoning model for financial forecasting. *Information Management & Computer Security, 13*(5), 394–405. https://doi.org/10.1108/09685220510582638
10. Klein, G. (2015). A naturalistic decision-making perspective on studying intuitive decision making. *Journal of Applied Research in Memory and Cognition, 4*(3), 164–168. https://doi.org/10.1016/j.jarmac.2015.07.001
11. Kubickova, A., Dohnal, M., & Doubravský, K. (2013). Qualitative decision-making model of investment into start-up companies. *International Journal of Technology Intelligence and Planning, 9*(1), 34–52. https://doi.org/10.1504/IJTIP.2013.059656

12. Lee, H., Kim, J., & Park, S. (2025, February 20). ReVISE: Learning to refine at test-time via intrinsic self-verification. *arXiv*. https://arxiv.org/abs/2502.07896
13. Ma, X., Wang, Y., & Liu, Z. (2025, February 18). S²R: Teaching LLMs to self-verify and self-correct via reinforcement learning. *arXiv*. https://arxiv.org/abs/2502.07184
14. McAndrew, C., & Gore, J. (2009). Day traders, computers, and the trading floor: Interactional expertise? *Proceedings of the European Workshop on Naturalistic Decision Making*, 34–40. https://doi.org/10.14236/EWIC/NDM2009.34
15. Min, S., Lee, H., & Choi, J. (2025). Self-regulation and requesting interventions. *arXiv*. https://arxiv.org/abs/2502.06975
16. Minsky, M. (1986). *The Society of Mind*. Simon & Schuster.
17. Olsen, R. A. (2002). Professional investors as naturalistic decision makers: Evidence and market implications. *The Journal of Psychology and Financial Markets*, *3*(3), 161–167. https://doi.org/10.1207/S15327760JPFM0303_4
18. Pink, S., Brown, T., & Williams, R. (2025). Position: Episodic memory is the missing piece for long-term LLM agents. *arXiv*. https://arxiv.org/abs/2502.06975
19. Pliske, R. M., & Klein, G. (2011). The naturalistic decision-making perspective. In G. Klein, B. Moon, & R. R. Hoffman (Eds.), *Linking Expertise and Naturalistic Decision Making* (pp. 39–56). Psychology Press. https://doi.org/10.1017/CBO9780511609978.019
20. Schraagen, J. M., Militello, L. G., Ormerod, T. C., & Lipshitz, R. (2012). *Naturalistic Decision Making and Macrocognition*. CRC Press.
21. Sprenger, J. (2017). Financial consulting: A qualitative study on its role in financial decision-making. *International Journal of Financial Studies*, *5*(1), 1–17. https://doi.org/10.4419/86788777
22. Xiao, L., Wang, Y., & Li, J. (2024, December 28). TradingAgents: Multi-Agents LLM Financial Trading Framework. *arXiv*. https://arxiv.org/abs/2412.12345
23. Wang, S., Li, X., & Zhang, Y. (2024, December 5). Reinforcement learning enhanced LLMs: A survey. *arXiv*. https://arxiv.org/abs/2501.01457
24. Wei, J., Chen, H., & Zhou, L. (2025, February 25). SWE-RL: Advancing LLM reasoning via reinforcement learning on open software evolution. *arXiv*. https://arxiv.org/abs/2502.08947
25. Yang, D., Zeng, L., Chen, K., & Zhang, Y. (2024, December 31). Reinforcing thinking through reasoning-enhanced reward models. *arXiv*. https://arxiv.org/abs/2501.05673
26. Yang, L. (2024). An empirical review of behavioral decision-making in the financial markets. *Advances in Economics, Management and Political Sciences*, *5*(2), 198–214. https://doi.org/10.54254/2754-1169/78/20241654
27. Yu, S. J., Park, S. C., & Wang, J.-C. (2001). Decision making using time-dependent knowledge: Knowledge augmentation using qualitative reasoning. *International Journal of Intelligent Systems in Accounting, Finance & Management*, *10*(2), 125–142. https://doi.org/10.1002/isaf.194

28. Ji, Y., Massanari, R. M., Ager, J., Yen, J., Miller, R. E., & Ying, H. (2007). A fuzzy logic–based computational recognition-primed decision model. *Information Sciences*, *177*(20), 4534–4550. https://doi.org/10.1016/j.ins.2007.04.001
29. Hu, Y., Li, R., & Zhang, Y. (2018). Predicting pilot behavior during midair encounters using recognition-primed decision model. *Information Sciences*, *430–431*, 373–386. https://doi.org/10.1016/j.ins.2017.11.015
30. Norling, E., Sonenberg, L., & Rönnquist, R. (2000). Enhancing multi-agent based simulation with human-like decision making strategies. In *Proceedings of the 1st International Workshop on Multi-Agent Based Simulation* (pp. 123–137). Springer. https://doi.org/10.1007/3-540-44561-7_9

Chapter 7

AI and Finance: Redefining Risk Analysis through Advanced Technologies

Ilia Munerman

Introduction

In today's rapidly evolving business landscape, the ability to accurately assess various risks – financial, fraud, or operational – is more crucial than ever. Financial scoring models traditionally used by banks to evaluate creditworthiness are just one part of a broader spectrum. Today, these models are being transformed to meet a wider range of needs, including fraud detection and operational risk assessment. Integrating generative artificial intelligence (AI) into these models revolutionizes how data is analysed and interpreted, offering deeper insights and more actionable information [1].

Clients of information services frequently require not just an evaluation of multiple coefficients but also a conclusive judgment delivered as a linguistic output. This enhances the utility of the data, making it accessible and actionable for decision-makers who may not have a deep background in data science [2].

Our case investigates the implementation of generative artificial intelligence (AI) technologies to enhance scoring systems used in different sectors of the national economy [3]. We delve into creating an algorithm that utilizes

 DOI: 10.4324/9781003606956-7

fine-tuning and prompting approaches to interpret scoring assessments more effectively. The algorithm is trained on high-quality financial analysis texts, enabling it to generate explanations for scoring decisions that are accessible and informative for clients. This use of generative AI not only improves the transparency of scoring models, which are often perceived as opaque "black boxes" by the public, but enhances communication and interactions with clients.

The article emphasizes the importance of continuous training and calibration of the model to adapt to dynamic market conditions, thereby reducing the risk of manipulation by malicious entities [4]. Integrating generative AI in scoring systems is a significant step towards more responsive and understandable financial services, aligning with modern demands for accountability and client engagement in financial assessments.

Overview of the Scoring Model

The scoring model is at the core of modern risk analysis, a sophisticated framework that evaluates various indicators to determine a company's risk across multiple dimensions. These models analyze data points such as revenue, net profit margins, debt levels, liquidity ratios, and fraud indicators to create a comprehensive picture of health and stability. Traditionally perceived as "black boxes," these models often produce outputs difficult for non-experts to understand. To bridge this gap, there has been a significant shift towards enhancing these models with AI technologies that improve not only accuracy but also the interpretability of the data [1].

Algorithmic Enhancements with AI

The "Textual Financial Analysis Algorithm" exemplifies the advanced integration of AI into traditional scoring models. This algorithm enhances the existing models by incorporating steps that systematically process financial indicators [4]. Initially, it selects key financial metrics, such as revenue growth or debt levels, and compares them with historical data to identify significant trends or anomalies. The algorithm calculates the percentage change for each indicator and benchmarks these against industry and regional averages [4] to create groups of data points that highlight areas of strength and concern [2, 5].

This structured approach quantifies financial health and prepares the data for further linguistic processing, making the insights gleaned from the analysis more accessible to decision-makers [6]. By automating the detection of significant financial trends and potential risk areas, the algorithm significantly speeds up the assessment process, allowing for more timely and informed decisions.

Integration with Generative AI

Integrating generative AI into scoring models is pivotal in making complex financial data understandable and actionable. Generative AI uses natural language processing (NLP) techniques to translate the scoring models' numerical data and statistical outputs into coherent, easily interpretable text [1]. This capability is crucial for providing clients with raw data and narrative reports explaining what the data means practically.

For instance, if a model identifies a significant deviation in the debt-to-income ratio, the AI can generate a narrative explaining the potential implications of this trend for the company's financial stability. This is done by leveraging a set of predefined phrases and contextual templates stored in an extensive database, such as the phrases provided in the manual-created financial analyses [7]. These templates are dynamically selected and combined based on the specific conditions and outcomes identified during the analysis, ensuring that each report is accurate and uniquely tailored to the situation.

By using linguistic outputs, these enhanced models can communicate complex financial concepts in a manner that is easy to understand, thus empowering clients to make better-informed decisions based on comprehensive and digestible information.

Case Study Examples

To illustrate how SPARK-Interfax implements textual financial analysis, consider how it transforms raw financial indicators into plain-language explanations. The system ingests a company's financial data (revenues, profits, margins, debt ratios, etc.) and automatically generates commentary highlighting key insights. This process involves several steps that bridge numeric data and natural language:

1. **Data Analysis**: The platform's scoring model first computes significant changes or ratios from the raw statements. It identifies notable trends (e.g., a sharp rise or fall in revenue, profit margin changes, liquidity ratios) and flags outliers or significant deviations from industry benchmarks.
2. **Phrase Selection**: The system selects an appropriate linguistic template or sentence for each key indicator. These templates are crafted by financial analysts and linguists so that each numeric pattern (such as "+20% revenue growth" or "profit turned to loss") maps to a precise explanatory phrase. For example, a revenue increase above a certain threshold might trigger phrases like "revenues grew significantly, indicating robust expansion." In contrast, a negative net income might trigger "the company incurred a net loss, raising concerns about sustainability."
3. **Narrative Construction**: The system weaves these individual sentences into a coherent narrative. The sentences are ordered and slightly adjusted for readability, producing a paragraph that reads as if written by a financial analyst. This narrative contextualizes the data, often combining multiple points into one explanation for conciseness.

Sample Output: To see this in action, suppose Company ABC's revenue grew from ₱80 million to ₱100 million (a 25% increase) last year, and its net profit margin improved from 5% to 8%. The SPARK-Interfax textual analysis module might produce a summary like:

> The company's revenue increased by 25% year-over-year, indicating a strong growth momentum in sales. Net profit margin also rose to 8% (up from 5% a year prior), reflecting improved cost efficiency and profitability.

This sentence pair takes the raw figures and explains their significance in a client-friendly way. Now imagine that in the same report, ABC's debt-to-equity ratio climbed and cash reserves fell. The system would add language to flag those issues:

> However, the debt-to-equity ratio doubled to 2.0, which may signal higher financial risk if earnings don't keep pace. The company's cash reserves also declined, potentially tightening its short-term liquidity.

SPARK-Interfax has converted numeric deltas and ratios into concise English (or Russian) commentary in these examples. A client reading the report can immediately grasp what's going well or poorly without poring over tables of numbers. The generative AI component ensures that each report is tailored to the company's actual performance – for instance, emphasizing growth in one case or warning of rising risks in another – all in natural language. This level of automated narrative was traditionally hard-coded or written by analysts. Still, SPARK's integration of language generation means the explanations are produced instantly for each company's unique data profile. The output is not just a generic template but a context-aware explanation (e.g., noting "robust expansion" vs. "raising concerns") that enhances clarity and usability for the client.

The linguistic engine behind this works with a library of curated phrases (and in some cases, AI-trained models) that ensure the tone remains professional and the content remains accurate. For instance, SPARK's library might define a 10–20% revenue growth as "moderate" while anything above 20% is "significant" or "strong," and the text will reflect that choice of wording. Likewise, if a company's profit turned positive to negative, the system might include a cautionary sentence about losses. All these rules and AI-generated variations result in a case-by-case "story" of the company's financial health, delivered alongside the complex numbers. This detailed example shows how raw indicators are translated into narrative insights, greatly simplifying financial analysis.

Enhanced Algorithmic Description: Fine-Tuning and Prompting in Textual Financial Analysis

The SPARK-Interfax system employs advanced AI techniques, specifically fine-tuning and prompting, to enhance the effectiveness and accuracy of its financial scoring models [1]. These methods are pivotal in transforming complex financial data into user-friendly narrative reports.

Fine-Tuning:

Fine-tuning in the SPARK-Interfax system involves adjusting a pre-trained generative model to better suit the specific nuances of financial language and the particular analytical needs of financial scoring. This process starts with a base model that has been pre-trained on a vast corpus of text encompassing a broad range of topics and styles. The fine-tuning phase involves re-training this model on a specialized dataset comprising financial reports, analysis texts, and historical financial data annotations [7].

This re-training adjusts the model's parameters so that it becomes more adept at recognizing and generating text patterns that are commonly found in financial contexts. For example, the model learns to detect significant financial terms and their relationships, such as "liquidity ratios," "net profit margins," and "debt levels," and understands how these terms typically interact within financial discourse.

Prompting

Prompting, on the other hand, involves creating specific queries or cues that guide the fine-tuned model in generating text directly relevant to the input data it analyses. In the SPARK-Interfax context, prompts are carefully designed to elicit the most informative and relevant narrative output based on the financial data provided.

For instance, if the model analyses a company's quarterly financial report, the prompt might be structured as follows: "Explain the significance of a 15% increase in net profit margin in the context of the company's year-over-year performance." This prompt directs the AI to focus its narrative on the implications of the observed profit increase, leveraging its trained knowledge base to generate a coherent and contextually appropriate analysis.

The combination of fine-tuning and prompting ensures that the AI not only generates text that is factually accurate but also contextually nuanced. The model can thus produce narrative analyses that feel natural and informative, similar to those a human financial analyst might write.

Implementation in Financial Analysis

By implementing these techniques, the SPARK-Interfax system effectively enhances its scoring models, making them capable of delivering narrative reports that accurately reflect the underlying data while being easy to understand. The fine-tuned and prompted AI can sift through extensive financial data, identify key trends and metrics, and articulate these findings clearly and concisely [4]. This capability is invaluable in helping clients quickly grasp complex financial situations and make informed decisions based on comprehensive, AI-driven insights.

Addressing Risks and Technical Challenges in AI-Driven Financial Analysis

While the integration of fine-tuning and prompting significantly enhances the SPARK-Interfax system's capability to generate accessible financial analyses, it also introduces certain risks and complexities that must be carefully managed.

Risk of Exposing Feature Space

One of the inherent risks in providing detailed explanations of financial scores, especially fraud scores, is the potential exposure of the feature space used in the models. By detailing how certain scores are derived, there is a possibility that fraudsters could gain insights into the aspects of financial behavior that the system monitors most closely [7]. This knowledge could enable malicious actors to tailor their activities to circumvent detection, exploiting known parameters to skew results in their favor. It is crucial to balance the level of detail provided in AI-generated explanations to mitigate this risk. The system can be designed to offer sufficient insight for legitimate users to understand their scores without revealing specific weights or thresholds that fraudsters could manipulate [8].

Challenges with Explaining Complex Dependencies

Another significant challenge arises when dealing with the complexities of explaining models that use advanced techniques such as fuzzy logic. Fuzzy logic, often employed in scoring models to handle uncertainties and approximate reasoning, can create non-straightforward or linear dependencies. Explaining the outcomes of such models in a way that is both accurate and accessible to non-experts can be particularly challenging. The AI system needs to be sophisticated enough to interpret these complex relationships and translate them into explanations that accurately reflect the underlying logic without oversimplifying the nuances.

To address these challenges, the SPARK-Interfax system incorporates several strategies:

- **Layered Explanations:** These offer explanations at different levels of detail, allowing users to delve deeper based on their expertise and need for information. However, high-level explanations are kept more general to avoid disclosing sensitive model specifics.
- **Advanced Natural Language Generation (NLG) Techniques:** Utilizing state-of-the-art NLG capabilities to articulate complex model behaviors in simpler terms. This involves leveraging context-aware generation techniques to interpret and explain nuanced logical relationships as part of the narrative output.
- **Continuous Monitoring and Updating:** Regularly updating the explanation models to reflect new patterns and emerging fraud techniques, ensuring that explanations remain relevant and do not inadvertently provide loopholes that could be exploited.

By implementing these strategies, the SPARK-Interfax system not only enhances the interpretability of its financial scoring models but also maintains the integrity and security necessary to protect against potential manipulation and abuse [9].

Benefits and Challenges of Integrative Generative AI in Financial Scoring

Integrating generative AI into credit scoring models (like the one in SPARK-Interfax) brings clear benefits to transparency and client understanding, but it also introduces new challenges in accuracy and governance. Below we discuss some key advantages of this approach, as well as important challenges and how they are being addressed:

Benefits

Enhanced Transparency and Explainability: Generative AI transforms complex financial data into understandable narratives, making scoring models more transparent. This clarity helps clients and decision-makers grasp the reasons behind their financial assessments, fostering trust and confidence in the analysis provided.

Improved Client Interactions: By automating the generation of narrative reports, these models facilitate communication with clients, who may not have the expertise to decode complex financial data. This enhances customer satisfaction and enables clients to make informed decisions based on clear, concise financial summaries.

Increased Efficiency: Integrating AI into financial analysis reduces the time analysts spend on routine data processing and report generation. This shift allows them to focus on more strategic tasks, such as interpreting complex scenarios or developing new financial products.

Consistency Across Reports: AI-driven models ensure the analysis remains consistent across different reports and clients. This uniformity is crucial for maintaining the reliability of financial assessments and reducing the likelihood of human error.

Challenges

Accuracy of AI-Generated Content: One of the primary challenges is ensuring the accuracy of AI-generated content. Misinterpretations or errors in data processing can lead to misleading conclusions, potentially affecting financial decisions. Regular updates and checks are necessary to maintain the integrity of AI systems [10].

Handling Sensitive Data: These systems process a significant amount of sensitive financial data, so maintaining privacy and security is paramount. Rigorous data protection measures must be in place to prevent unauthorized access and ensure compliance with global data protection regulations.

Adapting to New Financial Scenarios: The financial world is dynamic, with new products and financial instruments constantly emerging. The AI system must be adaptable and continually updated to analyse and interpret these new scenarios accurately.

Regulatory Compliance: Financial institutions are subject to stringent regulatory requirements. Ensuring AI-generated reports comply with these regulations is crucial to avoid legal repercussions and maintain operational legitimacy.

Conclusion

The integration of generative artificial intelligence into financial scoring models, as exemplified by the SPARK-Interfax system, represents a significant advancement in financial analysis. By transforming raw financial data into intelligible, narrative-based reports, these AI-enhanced models provide clients and decision-makers with accessible and actionable insights, demystifying financial assessments' often opaque nature.

Key Takeaways

- **Transparency and Understandability:** Generative AI makes financial evaluations more transparent and understandable, bridging the gap between complex data analysis and business decision-making.
- **Efficiency and Consistency:** Automating narrative report generation speeds up the analytical process and ensures consistency across different assessments, enhancing the reliability of financial evaluations.
- **Enhanced Client Interaction:** AI-driven models improve client interactions by providing clear and concise financial summaries, fostering a better understanding and facilitating informed decision-making.

Future Outlook

As the financial sector continues to evolve, the role of AI in economic analysis is expected to expand further. Future developments may focus on enhancing AI systems' adaptability to new financial products and market conditions, improving the precision of AI-generated content, and ensuring that these systems meet the increasingly stringent regulatory compliance requirements.

However, as we harness AI's benefits, we must also be vigilant about its challenges, such as maintaining accuracy, ensuring security, and adhering to regulatory standards. Ongoing research, development, and regulation will be critical in shaping AI's future in financial analysis, ensuring that it continues to provide value while mitigating risks.

In conclusion, while generative AI presents a promising tool for financial analysis, its successful integration requires a balanced approach that considers its potential and pitfalls. By continuing to develop robust AI solutions and addressing the challenges head-on, the financial industry can look forward to a future where financial assessments are more accessible, insightful, and influential than ever before.

References

1. Hastie, T., Tibshirani, R., & Friedman, J. (2009). *The elements of statistical learning: Data mining, inference, and prediction* (2nd ed.). Springer-Verlag. ISBN: 978-0-387-84857-0. https://hastie.su.domains/ElemStatLearn/
2. Piotroski, J. D. (2002). *Value investing: The use of historical financial statement information to separate winners from losers* (Selected Paper No. 84). The University of Chicago Graduate School of Business. Retrieved from http://www.chicagobooth.edu/faculty/selectedpapers/sp84.pdf
3. Khoso, A., & Yusof, A. (2019). Extended review of contractor selection in construction projects. *Canadian Journal of Civil Engineering, 47.* https://doi.org/10.1139/cjce-2019-0258
4. Pesaran, M. H., Schuermann, T., & Weiner, S. M. (2004). Modeling regional interdependencies using a global error-correcting macroeconometric model. *Journal of Business and Economic Statistics, 22,* 129–162.
5. Klerck, W. G., & Maritz, A. C. (1997). A test of Graham's stock selection criteria on industrial shares traded on the JSE. *Investment Analysts Journal, 45.* Retrieved from http://www.iassa.co.za/articles/045_1997_03.pdf
6. Rousseeuw, P. J. (1987). Silhouettes: A graphical aid to the interpretation and validation of cluster analysis. *Journal of Computational and Applied Mathematics, 20,* 53–65. https://doi.org/10.1016/0377-0427(87)90125-7
7. Shibani, A., Hassan, D., Saaifan, J., Sabboubeh, H., Eltaip, M., Saïdani, M., & Gherbal, N. (2022). Financial risks management within the construction projects. *Journal of King Saud University – Engineering Sciences.* https://doi.org/10.1016/j.jksues.2022.05.001
8. Spence, M. (1981). Signaling, screening, and information. In S. Rosen (Ed.), *Studies in labor markets* (pp. 319–358). University of Chicago Press. ISBN: 0-226-72628-2.
9. Tarawneh, S., & Kasabreh, N. (2019). Investigating the impact of contractor's performance on the success of Jordanian residential construction projects. *International Journal of Construction Management, 21.* https://doi.org/10.1080/15623599.2018.1560547
10. Witten, I. H., & Frank, E. (2005). *Data mining: Practical machine learning tools and techniques* (2nd ed.). Morgan Kaufmann. ISBN: 978-0-12-088407-0. https://www.cs.waikato.ac.nz/ml/weka/book.html

Chapter 8

AI in Financial Fraud Detection and Prevention

Shaik Valli Haseena, Neha Jaswani, Ayasha,
Gaikwad Beena Suresh, and Simna Shanavas

Introduction

The advent of digital transactions has made financial fraud more intricate and frequent, creating major obstacles for global financial institutions. Traditional rule-based fraud detection systems often fall short in identifying novel and sophisticated fraudulent activities. Artificial intelligence (AI) stands as a transformative solution that delivers sophisticated methods to effectively identify and thwart financial fraud.

Machine learning algorithms empower AI-driven systems to examine extensive datasets where they detect patterns and anomalies that suggest fraudulent behaviour. These systems develop their detection abilities by adjusting to new fraud techniques through continuous evolution. Neural networks and other deep learning models handle intricate unstructured datasets to detect subtle fraud indicators that traditional methods fail to identify.

Fraud detection systems receive additional support through Natural Language Processing (NLP) techniques which examine communication texts to identify phishing attempts and scam messages. AI-driven real-time anomaly detection systems scrutinize ongoing transactions to deliver instant responses to detected irregular activities. The examination of user

DOI: 10.4324/9781003606956-8

interactions through behavioural biometrics involves analyzing typing patterns and mouse movements to confirm identities while detecting unauthorized access.

Artificial intelligence when applied to financial fraud detection systems boosts both the precision and velocity of fraud identification while simultaneously strengthening financial system security. The evolution of fraudster techniques demands that AI systems constantly learn and adapt to sustain effective financial crime defences.

Components of AI in Financial Fraud Detection

Artificial intelligence has revolutionized the landscape of financial fraud detection by enabling institutions to analyze vast amounts of data and identify suspicious activities with greater speed and accuracy than ever before. By leveraging advanced machine learning techniques, AI systems help banks and other organizations adapt to rapidly evolving fraud schemes and proactively protect against financial crime.

Machine Learning Algorithms

Machine Learning (ML) algorithms are responsible for scanning enormous volumes of transactional data to identify and prevent fraudulent transactions. ML algorithms allow banks and other financial institutions to identify sophisticated fraud patterns that rule-based systems fail to recognize. Through [1] the use of AI-based methods, fraud detection models can refine their accuracy with time and respond to changing fraudulent patterns. Fraud detection using ML is largely reliant on three broad categories: supervised learning, unsupervised learning, and reinforcement learning.

Supervised Learning models are founded on labelled data, which label historical transaction data as fraud or genuine. The models get trained on past fraud patterns and use the learning to detect fraud in real time. Methods like Logistic Regression, Random Forest, and Gradient Boosting Machines (GBMs) [2] are used extensively for detecting fraud. The biggest advantage of supervised learning is its accuracy in detecting well-established fraud patterns; however, its performance relies on the quality and amount of available labelled data.

Unsupervised Learning does not require prelabeled data and is therefore ideal for detecting new and evolving patterns of fraud. The algorithms process transaction behaviour and choose patterns that are not in line with standard user behaviour. K-means and DBSCAN are popular clustering algorithms used in fraud detection, and Autoencoders and Principal Component Analysis (PCA) [3] are used as anomaly detection methods. Unsupervised learning works well at detecting unnamed patterns of fraud but produces more false positives than supervised methods.

Reinforcement Learning is an adaptive method of learning dynamically through interactions and updating fraud detection strategies over time. It uses reward-based feedback operations to modify the detection rules and improve decision-making. The method is especially beneficial in adaptive fraud detection systems where fraud methods dynamically change. The reinforcement learning models improve the capabilities of the system to counteract fraud in a proactive manner by dynamically changing the detection thresholds and improving the risk scores.

A practical example of a successful anomaly detection algorithm for fraud detection is the Isolation Forest. It identifies fraudulent transactions by recursively dividing the data, thereby efficiently identifying outliers with low computational overhead. In contrast to conventional classification models, the Isolation Forest operates without labelled data, and hence it is particularly well-suited for real-time fraud detection in dynamic financial scenarios.

By integrating these machine learning techniques, banks can create efficient fraud prevention systems that optimize financial losses, enhance transaction security, and enhance user trust. Employing multiple ML techniques offers a more robust fraud prevention system that responds to ever-changing threats in the digital financial market.

Neural Networks and Deep Learning

Deep learning technology has revolutionized the scene of fraud detection by enabling banks to analyze large and complex collections of data at a highly precise pace. Unlike what machine learning would otherwise accomplish, deep learning models can learn usable features without manual explicit specification, a fact that makes them highly effective for detecting complex fraud patterns. The most important deep learning architectures implemented for fraud detection include Convolutional Neural Networks (CNNs) and Recurrent Neural Networks (RNNs), the two of which display improved performance in handling different types of financial data.

Processing Sequential Data Using RNNs

Fraud behaviour is prone to exhibit temporal characteristics, with the suspicious patterns of transactions being exhibited over time as opposed to in a one-off fashion. Recurrent Neural Networks (RNNs) [4] are designed with sequence data in mind, and therefore they are highly capable of identifying time-based anomalies. By processing a sequence of transactions, RNNs can identify anomalies in user expenditure patterns, infrequent frequencies of transactions, or purchases that are out of geographic context. Improved variants like Long Short-Term Memory (LSTM) networks greatly enhance fraud detection by identifying long-term dependencies, thus ensuring that past fraud signals are considered in identifying suspicious transactions.

Processing Unstructured Data with CNNs

Fraud detection has long been based on structured numerical data; however, Convolutional Neural Networks (CNNs) [5] are at the forefront of the analysis of unstructured forms of data, including scanned documents, images, and transaction receipts. The financial fraud examples are liable to include activities such as document forgery, identity theft, or invoice manipulation. CNNs exhibit great capability in detecting subtle image changes and in finding fraudulent documents. Through the application of deep feature extraction, CNNs significantly improve fraud detection systems by authenticating document integrity, identifying stamp anomalies, and finding exploited financial records.

Managing Non-Linear Relationships

One of the most significant strengths of deep learning models is that they can identify sophisticated, non-linear patterns in financial data. Most conventionally used fraud detection systems are rule-based and feature engineering-based models, which can be blind to underlying fraud patterns. Deep learning models, however, learn sophisticated interactions between multiple variables from data without any human intervention, which boosts detection rates. By combining CNNs and RNNs, financial institutions can create a strong fraud detection system that identifies transactional anomalies and document-based anomalies. By using deep learning methods, fraud detection systems become more scalable, flexible, and resistant to changing fraud strategies. The models not only minimize false positives but also improve real-time fraud prevention, providing greater security in financial transactions on banking, e-commerce, and fintech platforms.

The architecture of a typical deep learning model for fraud detection can be visualized as in Figure 8.1.

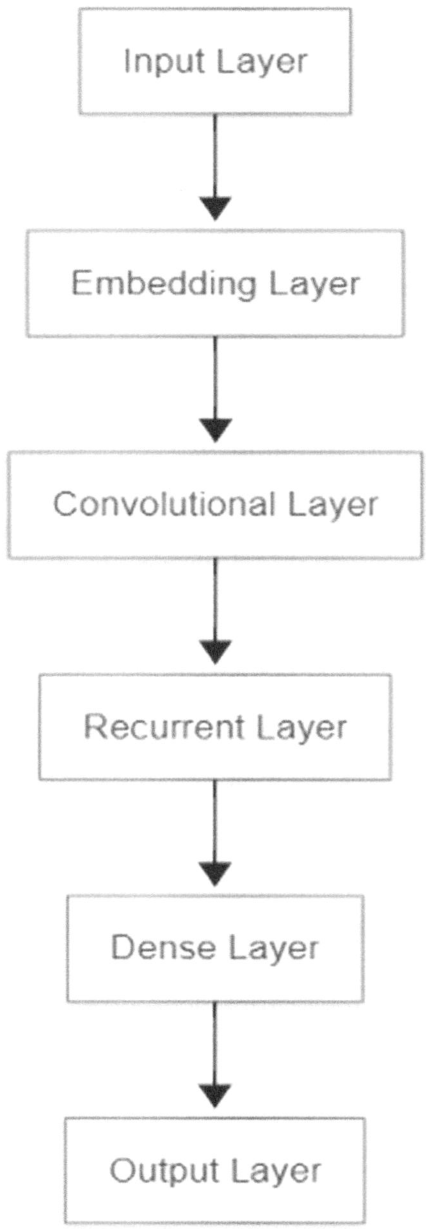

Figure 8.1 Fraud detection system.

The image illustrates a deep learning fraud detection system, which is at the core of identifying suspicious financial transactions. The model architecture is able to efficiently process large amounts of transaction data, enabling real-time fraud detection and prevention. All the layers of the model contribute to feature extraction, learning patterns, and making accurate fraud classification.

The Input Layer is the base of the model, where raw financial transaction data is fed in. The data comprises transaction amounts, time stamps, locations of users, payment modes, and user behaviour patterns. As financial fraud typically lies concealed in intricate patterns of transactions, the model has to preprocess and organize the data in a suitable form for subsequent analysis.

The Embedding Layer transforms categorical features, such as merchant IDs, user account numbers, and transaction types, into dense numerical vectors. This allows the model to learn latent correlations among different entities in transactions, thus making it capable of discovering subtle frauds. Through the learning of distributed representations of the data, this layer enhances feature extraction and improves the overall performance of the model.

The Convolutional Layer is responsible for detecting fraud patterns that are localized. It applies convolutional filtering to detect anomalies such as unusual spikes in high-value transactions, repeated failed payments, or unusual spending behaviour. It helps to extract spatial patterns from transactions, leading to higher accuracy in detecting fraud.

The Recurrent Layer is used to identify sequence dependencies and thus is very powerful in analyzing past patterns of transactions over time. Most financial fraud is time-based behaviour, i.e., repeated unauthorized withdrawals from various locations within a short duration window. By the use of models such as Long Short-Term Memory (LSTM) or Gated Recurrent Units (GRU) [6], the layer eases the identification of continual patterns of fraud and reduces the false negatives.

The Dense Layer integrates features learned from prompt layers and assigns weights to multiple indicators of fraudulent behaviour. Fully connected neurons operate on high-dimensional transaction data to improve fraud classification. This layer enriches decision-making by integrating data from diverse sources, thus enabling the model to differentiate between legitimate and fraudulent behaviour with high accuracy.

The Output Layer is responsible for producing the conclusive classification results regarding fraud. It generally utilizes activation functions like softmax or Sigmoid to generate either binary outcomes (indicative of fraudulent or legitimate behaviour) or multi-class fraud predictions. Subsequently, this output is leveraged to activate real-time fraud prevention strategies, which may include notifying financial institutions, obstructing dubious transactions, or commencing multi-factor authentication processes for activities assessed as high risk. This deep learning architecture powered by AI significantly improves detecting financial fraud through enhanced detection accuracy, minimizing false positives, and evolving to changing fraud tactics. Its use across banking, e-commerce, and fintech markets enhances financial security, lowers economic losses, and provides a more secure transaction environment.

Natural Language Processing (NLP)

Natural Language Processing (NLP) techniques have proved to be a useful tool in the detection of fraudulent intent within text-based messages, especially within financial services and transactions. With the processing of large text datasets, NLP is able to identify minute indications of fraud that may otherwise go undetected by traditional means. Among the range of NLP techniques, the three that proved especially useful for fraud detection are Sentiment Analysis, Keyword Detection, and Entity Recognition.

Sentiment Analysis is probably the most related method in NLP employed to detect fraudulent intent. It entails the analysis of the emotional tone or sentiment of a message to determine if the communication seems suspicious or manipulative. For instance, scammers employ extremely positive or urgent wording to elicit an emotional reaction, such as evoking a sense of urgency or offering unrealistic benefits. By examining the sentiment of the message, financial institutions can detect these subtle signals and mark messages that could be a scam. For instance, an email with words such as "act now" or "limited time offer" could be marked for inspection, as these are typically linked to fraudulent schemes intended to coerce people into making quick decisions.

The detection of fraudulent messages through Keyword Detection in the context of Natural Language Processing (NLP) has become a major area of research in cybersecurity and fraud prevention. Early approaches to fraud detection relied largely on rule-based filtering techniques that manually picked up suspicious phrases. However, these methods suffered from limitations in flexibility and struggled to detect new patterns of fraud. As artificial intelligence and NLP developed, modern fraud detection systems today leverage machine learning algorithms in combination with contextual analysis to detect ever-changing tactics used by fraudsters. Studies have shown that scammers commonly use language that creates urgency to trick victims, such as with phrases like "your account has been hacked" or "urgent transaction needed." Research on phishing detection has shown that the detection of high-risk words, combined with sentiment analysis and entity detection, greatly upgrade the accuracy in the detection of guileful communications.

Deep learning focused NLP models like transformers and RNNs have been used to find not only preset keywords but also deviations of fraudulent phrases in real time. The combination of keyword detection with anomaly detection paradigms has also hardened fraud prevention efforts, allowing financial institutions to automatically scan, classify, and flag suspicious messages before any harm is caused. The fraudsters continue to evolve their approaches, and the development of more adaptive and context-aware NLP-based fraud detection models remains a critical area for research.

Entity Recognition, or Named Entity Recognition (NER) [7], is the identification and validation of certain entities that are referred to in a communication. These entities may be personal names, banks, account numbers, and other critical information. For example, in a phishing email, an attacker may fake a legitimate bank's name or name a reputable company to trick the victim. By using NER methods, the system can identify the validity of the entities referred to in the message and cross reference them with known databases. If the entities are suspicious or do not correlate to known institutions, the communication can be flagged for closer inspection. This process prevents fraudsters from impersonating organizations or individuals in their efforts to trick victims.

Artificial intelligence-based financial transaction monitoring systems apply Natural Language Processing (NLP) techniques to improve the detection and prevention of fraud using transactional data analytics and communication patterns between clients and financial institutions.

The systems are far more sophisticated than traditional fraud detection systems by scrutinizing text communications, detecting suspicious messages, and identifying potential fraudulent activities. For instance, if a client continues to receive unwelcome messages requesting confidential information, or if a message contains high-risk words and unsettling emotions, the system can raise real-time alerts, thereby preventing potential threats before they are at an advanced stage. This proactive approach in fraudulence prohibition immensely enhances the system's capability to detect fraudulent activities at an early stage, making the AI-based fraud detection honed, more adaptable, and effective.

Besides, integration of NLP techniques such as Sentiment Analysis, Keyword Detection, and Entity Recognition enables financial institutions to better identify fraudulent or malicious messages. By analyzing text data for discrepancy, these AI systems can uncover hidden fraud patterns, identify intent, and prevent financial exploitation, thus enhancing security and reliance in digital transactions. Because fraudsters are constantly evolving their plan, the creation of NLP-based fraud detection remains crucial in maintaining robust cybersecurity measures in financial settings.

Anomaly Detection Systems

Artificial intelligence-based anomaly detection systems have proven to be essential in the war against financial fraud, using complex algorithms to review transactions in real-time and flag suspicious activity. These systems function by continuously assessing transactional information and comparing it to preset behavioural baselines for specific accounts or users. Statistical modelling is a normal method used in the area of anomaly detection, particularly for identifying outliers in assignable data. Analytical modelling utilizes mathematical and analytical ways to create a model of normal user or account behaviour through previous patterns of transactions. Following model training, any transaction outside the set statistical parameters like abrupt changes in the amount, location, or frequency of transaction will be marked as an outlier. For example, if a user habitually makes small local transactions, a large foreign transaction would be flagged as anomalous. Statistical modelling is efficient in identifying fraudulent transactions that are typified by sudden, odd behaviours, however, it may be inefficient in identifying more nuanced or sophisticated fraud patterns that do not create dramatic outliers.

Clustering algorithms add an added level of sophistication to the process of anomaly detection by grouping similar transactions based on shared attributes. Clustering algorithms use unsupervised learning techniques to identify patterns in large datasets in the absence of noted instances. By grouping transactions that exhibit similar behaviour such as transactions that happen within the same temporal window, geographic location, or transaction type, clustering algorithms are highly useful at identifying transactions that deviate from typical behaviour. For example, if a group of transactions constitutes normal spending habits, and a new transaction is outside of this group, it can be found as suspicious. This method is particularly useful when fraud closely mimics legitimate patterns of transactions, since the algorithm identifies deviations from the norm within the group.

Hybrid models mix the strengths of many detection methods to build the overall accuracy and effectiveness of fraud detection systems. Hybrid models combine heterogeneous approaches, statistical models, clustering algorithms, and machine learning-based organizers into one framework. Leveraging the strength of each technique, hybrid models can give more effective detection capacity while minimizing false positives. Statistical modelling, for example, can find large-scale agitation from expected behaviour, while clustering algorithms can find more covert rarity in batches of transactions. Combining the strategy provides a more complete examination of transactional data, enhancing the system's ability to detect overt and covert cases of fraud. Hybrid models are also very flexible, and financial institutions can tailor their fraud detection systems to the unique characteristics of their data and the dynamic techniques used by fraudsters.

AI-driven anomaly detection systems are a critical component of real-time fraud prevention since they continuously evaluate transactions and identify anomalies from normal behaviour. Different methodologies, such as Statistical Modelling, Clustering Algorithms, and Hybrid Models, each make a distinct contribution to this function by offering different solutions for anomaly and fraud detection. Statistical models are good at identifying large outliers, whereas clustering algorithms show better performance in identifying slight deviations in groups of transactions.

An example architecture for real-time fraud detection is shown in Figure 8.2.

The diagram shows a straightforward process for fraud detection using AI systems. It shows the primary steps involved in detecting fraud. This systematic approach improves the accuracy of fraud detection and reduces unnecessary false positives.

The financial industry is being increasingly targeted by sophisticated fraud schemes, and hence AI-based fraud detection systems become a necessity for transaction security. The diagram in Figure 8.2 illustrates a clear flow that aids in effectively detecting and helping fraudulent activity.

It starts with Data Ingestion. Here, the financial information is acquired from different sources such as transaction logs, user profiles, and device fingerprints. This gives real-time as well as historical accounts of transactions to be analyzed. Since fraudsters employ so many variable techniques,

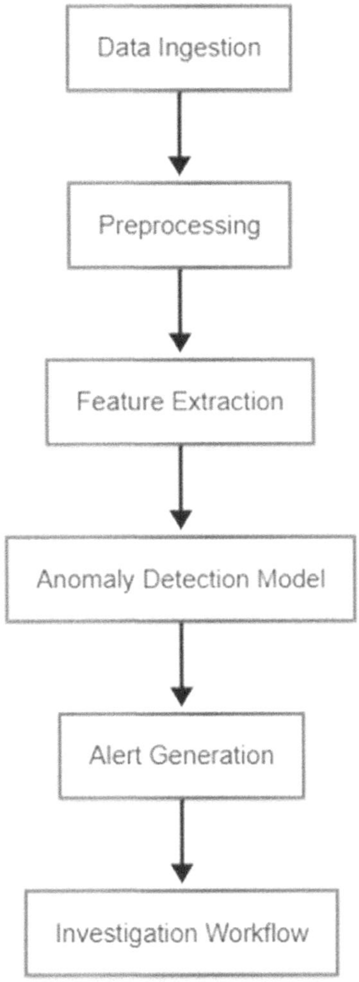

Figure 8.2 AI-driven fraud detection workflow.

it is important to collect so much data so that fraud can be identified appropriately.

In Preprocessing, raw data is converted to enhance its quality and utility. This involves handling missing values, normalizing transaction amounts, and converting categorical data to numerical data. Correct preprocessing ensures that AI models perform optimally and provide correct fraud predictions. The uprooted features enable machine learning models to distinguish normal and abnormal behaviour more correctly.

The Anomaly Detection Model is the core of the system. It utilizes AI methods like supervised learning, Random Forest, XGBoost, and unsupervised learning autoencoders, clustering algorithms to detect anomalies.

These models update themselves in real time to evolve with new fraud methods based on past fraud behaviour and changing detection parameters.

Notifications are started by the Alert Generation step once a legitimate scam has been found. AI-powered risk assessment classifies the warnings to remove false positives and only highlight high-risk transactions that need quick concentration. This reduces needless transaction disturbance and increases the effectiveness of fraud detection.

The examination workflow comprised reviewing commercial fraud that has been detected and taking appropriate action. Manual authentication, transaction blocking, or additional authentication may be used in high-risk circumstances. Continuous observation and feedback help to enhance fraud observation algorithms for improved system performance overall.

In order to prevent fraud threats, protect consumers, and ensure secure transactions, this machine learning-based fraud detection technology is extensively used in the banking e-commerce and fintech industries. By employing anomaly detection techniques and machine learning, banks may actively stop emerging fraud patterns and improve.

Implementation Challenges in E-commerce and Entertainment Industries

AI-powered fraud detection has become vital for e-commerce and entertainment businesses as both industries face rapidly growing digital threats. Innovative solutions must balance robust risk management with maintaining seamless and satisfying customer experiences in highly competitive online environments.

E-commerce Industry

The use of AI-based fraud detection in e-commerce has a number of challenges that need to be addressed through advanced solutions to efficiently manage risk while ensuring a positive customer experience. These challenges are mainly focused on the high volume of transactions, varied payment modes, and the problem of false positives. All these challenges need thoughtful consideration and adjustment of AI models to manage the nature and volume of e-commerce transactions.

High Transaction Volumes are among the biggest issues facing the e-commerce sector. With so many millions of transactions taking place on an average day, AI-based fraud detection systems have to be able to process huge volumes of data in real time in order to catch any fraudulent behaviour.

To process such large datasets effectively, AI systems tend to use distributed computing frameworks that can support parallel processing on numerous machines. These architectures, when combined with streaming algorithms in real time, ensure the detection system is scalable as well as fast. The solution to scaling AI solutions is possible due to the capability of models such as Random Forests, XGBoost, and Deep Neural Networks (DNNs) to parallelize data. Practically, these models can be trained for identifying typical patterns of fraud and keeping computational complexity low. The equation for a real-time fraud detection system could be:

$$\text{Fraud probability} = \text{Sigmoid}\ (w1.x1 + w2.x2 + \ldots + wn.xn) \tag{8.1}$$

where $x1$, $x2$, ..., xn are the input features (transaction amount, time, user history), $w1$, $w2$, ..., wn are the model-learned weights, and the Sigmoid function returns a probability between 0 and 1 representing the probability of a transaction being fraudulent. By implementing such models in a distributed setting, high volumes of transactions can be processed with little delay.

Variety of Payment Channels provides another dimension of complexity to fraud detection tools in online shopping. Customers employ a variety of payment channels, including credit cards, digital wallets, bank transfers, and cryptocurrencies. Each of these payment channels may exhibit unique patterns of fraud making it tough to create a one-size-fits-all model. For example, credit card fraud commonly includes unauthorized transactions from a remote geography, whereas digital wallet fraud can include the stealing of access credentials and quick money transfers. To combat this, e-commerce websites need to customize their fraud detection models to consider the specific attributes of each payment. A pragmatic strategy is the employment of multi-model architectures, in which different models are developed for every type of payment. These models can then be combined in a hybrid system to detect the fraud patterns unique to each payment method. A credit card fraud model could look for such factors as:

$$\text{Fraud detection} = f\,(\text{Transaction amount, Location distance, Previous spending patterns}) \tag{8.2}$$

where f is a machine learning function (Decision Tree, SVM) that computes the probability of fraud against the input features. For digital wallets, another model could be dedicated to device fingerprinting or login behaviour analysis, understanding that fraud in digital wallets tends to appear in the form of unauthorized logins from new devices or locations.

False positives are one of the most significant challenges in implementing AI-driven fraud detection systems in e-commerce. A system that is too sensitive and marks genuine transactions as fraudulent can cause customer dissatisfaction and loss of revenue. To counter this, fraud detection models need to be carefully calibrated to reduce negative positives without compromising on detecting fraudulent activity. This equilibrium is attained by the sensitivity and specificity of the model. Sensitivity is the capacity of the model to identify correctly fraudulent transactions, while specificity is the capacity of the model to identify correctly legitimate transactions.

The sensitivity and specificity are usually compared using the F1 score, which is the harmonic mean of recall (sensitivity) and precision. The $F1$ score formula is:

$$F1 = \frac{2.Precision.Recall}{Precision + Recall} \tag{8.3}$$

where

$$Precision = \frac{TP}{TP + FP}, Recall = \frac{TP}{TP + FN}$$

- TP = True Positives (fraudulent transactions correctly identified)
- FP = False Positives (legitimate transactions incorrectly flagged as fraudulent)
- FN = False Negatives (fraudulent transactions missed by the system)

An ideal AI system for fraud detection attempts to maintain a high $F1$ score with the lowest number of false positives (FP) but a high recall, so that fraudulent transactions will be flagged. In implementation, it incorporates methods like active learning and feedback loops in which the system continually improves on its predictions by observing the results of past fraud detection efforts.

The high transaction volume, multifaceted payment channels, and false positives inherent in e-commerce call for sensitive and scalable design and implementation of AI-based fraud detection systems. With real-time processing, custom models for individual payment channels and optimized algorithms for sensitivity trade-off e-commerce websites can safely mitigate the effect of fraud while preserving a user-friendly experience. Employment of mathematical equations like fraud probability and the $F1$ score offers a systematic approach to assess and optimize these AI models in order to improve fraud detection results constantly.

Entertainment Industry

The entertainment sector, especially in the context of streaming services, is confronted with a number of major fraud detection and security challenges. These include challenges around account sharing and credential theft, piracy of content, and data privacy issues. With the increasing demand for digital content, streaming services need to create advanced AI-based solutions to tackle these issues without compromising user experience or infringing on legal obligations. Sharing accounts and stealing credentials are common problems in the streaming sector.

Users tend to share their credentials with others, thus resulting in unauthorized access and revenue losses for the service providers. Additionally, credential theft is on the rise with hackers gaining access to users' accounts through phishing, data breaches, or brute force attacks. Detection of unauthorized access without causing inconvenience to legitimate users is a fine line. AI solutions can assist by monitoring patterns in user activity, including login frequency, geographical location of access requests, and the devices used for streaming content.

Through the use of anomaly detection methods, AI can detect abnormal behaviour, for example, an unexpected login from a new location or device, and mark these actions as possibly fraudulent.

A simple formula to describe the fraud probability of a login attempt could be:

$$\text{Fraud probability} = \text{Sigmoid}\,(w1.x1 + w2.x2 + \ldots + wn.xn) \tag{8.4}$$

where $x1$, $x2$, ..., xn are the features of the time location device and $w1$, $w2$, ..., wn are the model-learned weights. The Sigmoid function returns a probability value between 0 and 1 which indicates whether the login attempt is most likely to be fraudulent. To prevent inconveniencing honest users, such systems can be implemented to flag only high-risk behaviour, striking a balance between security and convenience to the user.

Content piracy is another significant issue, as pirated distribution of copyrighted material can result in financial loss for streaming providers. Piracy usually entails illegal sharing or selling of content on different online platforms, and it is challenging for providers to track and stop these activities. AI-based systems can be utilized to identify pirated content by applying digital watermarking, which inserts a special, invisible identifier into the content. The AI software can then search the web, social networking sites, and file-sharing sites for these watermarks, marking them automatically for unauthorized distribution. Machine learning programs can also study metadata, file types, and other technical parameters of shared content to detect potential cases of piracy, even where the watermark is removed or manipulated.

Employing Convolutional Neural Networks (CNNs) is another method used to detect piracy of content, especially of video and audio content. CNNs are able to analyze the patterns of audio and visual aspects, hence opt for identifying the pirated version by comparing them with original or licensed versions. This is done through feature extraction, where CNNs are trained to recognize minute differences between pirated and original copies like compression artefacts or frame changes. The models output might resemble:

$$\text{Piracy score} = f(\text{Feature1}, \text{Feature2}, \ldots, \text{FeatureN}) \tag{8.5}$$

where the characteristics are different facets of the content being analyzed, e.g., frame quality, compression ratio, or visual anomalies. A high piracy score would mean the content is most likely to be pirated.

Concerns regarding data privacy are always there, especially in the entertainment sector, which gathers enormous amounts of personal and behavioural data about users. AI-based anti-fraud systems need to be designed with specific care so that they comply with privacy laws like the General Data Protection Regulation (GDPR) in Europe or the California Consumer Privacy Act (CCPA) in the United States. This implies that streaming services will have to take care that AI models do not breach user privacy by unauthorized access or misuse of personal information. For instance, AI systems should refrain from processing users' personally identifiable information personally identifiable information (PII) without user consent, and they should incorporate mechanisms to anonymize user data where possible.

A real-world solution to these privacy issues is through differential privacy methods. Mathematically, the noise addition can be expressed as

$$\text{Perturbed data} = \text{Original data} + N(\mu, \sigma 2) \tag{8.6}$$

AI-based solutions in the entertainment sector need to be created to solve intricate issues like account sharing and credential theft, content piracy, and data privacy issues. Through the use of sophisticated AI methods like anomaly detection, machine learning, and CNNs, streaming platforms can identify fraudulent activity efficiently without compromising security and user experience. Moreover, by using privacy-preserving techniques like differential privacy, AI systems can be made compliant with data protection laws so that user information is processed responsibly without facilitating fraudulent activity. The application of these technologies allows streaming services to preserve the integrity of their platforms while protecting their content and user data.

AI in Financial Fraud Prevention

AI-driven fraud prevention represents a significant leap forward from traditional methods by enabling dynamic, automated responses to emerging threats. These technologies continuously learn from new data, allowing financial institutions and e-commerce platforms to stay ahead of increasingly sophisticated fraud tactics while minimizing disruption to legitimate customers.

Real-Time Fraud Prevention

Artificial intelligence-based fraud prevention systems have transformed the means by which e-commerce sites and financial institutions combat fraudulent transactions in real-time. Through real-time analysis of transactions, AI systems are able to identify suspicious behaviour and prevent threats before any damage is caused [8]. These systems use sophisticated machine learning algorithms to spot irregularities so they can react quickly to potential fraud. AI's main roles in fraud detection are to stop questionable transactions before they are completed, send out notifications for follow-up checks, and use multi-factor authentication to increase security and provide a robust defence against fraudulent transactions [9].

For example, if the large transaction is attempted from an unknown geographical location or if the mode of payment differs from the typical behaviour of the account owner, the system can automatically reject the transaction. The likelihood of fraud in this scenario can be modelled mathematically with a logistic regression function:

$$\textit{Fraud Probability} = \frac{1}{1+e^{-(w1.x1+w2.x2+\ldots+wn.xn)}} \quad (8.7)$$

where the attributes describing the transaction features (e.g., amount, time, location), are the model-learned weights during training. If the output probability is above a threshold, the system will reject the transaction. This method allows for rapid decision-making, such that fraudulent transactions are rejected prior to processing, reducing potential financial loss.

Triggering Alerts for Additional Verification is yet another important functionality of AI-based fraud prevention systems. Instead of blocking a transaction outright, AI systems have the ability to mark suspicious activity and trigger alerts for additional verification by the user or an operator. This mechanism finds a balance between security and convenience of usage, as it provides for the intervention to be made immediately when needed without interfering with genuine activity. Alerts could be triggered on the basis of unusual patterns in transactions, speed of access of accounts from diverse devices, or transactions occurring at unusual geographical areas. AI could be combined with a scoring method, where a risk score would be assigned to every transaction by measuring how deviant it was from the behaviour of the user. The model for determining the risk score for a transaction may be:

Risk Score is the sum of the product of weights and variable value changes, i.e., mathematically represented as

$$\sum_{i=1}^{n} w_i \cdot \Delta x_i \quad (8.8)$$

Here w_i is the weight or significance of each attribute, and $\Delta\, x_i$ is the difference between the attribute value (e.g., location, amount) and the user's history. A high-risk score would be the cause of an alert, requesting the user to verify their identity or authorize the transaction through a second security verification like two-factor authentication (2FA). Multi-Factor Authentication (MFA) is also a security mechanism employed by AI-based fraud detection systems. MFA requests users to authenticate through different authentication factors, thus rendering it difficult for fraudsters to gain access. Common practices entail the combination of knowledge-based factors (passwords), possession-based factors (security tokens or mobile phones), and inherent factors (intrinsic biometric data), e.g., fingerprints or facial recognition. Artificial intelligence can monitor login sessions and detect suspected fraudulent activity with invalid login credentials or out-of-pattern access patterns. For example, when a user logs in from a new location or device, the AI system can initiate an MFA request, thereby enabling security prior to accessing sensitive resources. The AI model can also assess the likelihood of fraud by merging various verification factors using a mathematical formula such as

$$\text{Fraud probability} = \text{Sigmoid}\,(w1\cdot\text{Password strength} + w2\cdot\text{Device recognition} + w3\cdot\text{Biometric match}) \tag{8.9}$$

Every attribute Password strength, Device identification, and Biometric match is assigned a separate weight ($w1$, $w2$, $w3$) to express its value in fraud detection. The Sigmoid function output will be a decision on whether the authentication is authentic or requires additional verification processes. This multifaceted process renders it even more challenging for fraudsters to bypass security controls, thus ensuring that sensitive financial accounts or data are accessed by authorized users only.

Artificial intelligence-based fraud detection and prevention systems offer real-time intervention and monitoring for fraud protection. Through preventing suspicious transactions before execution, triggering alarms for further scrutiny, and application of multi-factor authentication, such platforms provide end-to-end protection for online transactions. Leveraging advanced algorithms such as logistic regression and risk score models, such platforms can effectively assess the risk of a transaction and with the integration of multi-factor authentication provide a crucial defence against unauthorized transactions. Through continuous learning and adaptation, AI-powered fraud prevention platforms can improve their ability to detect and react to threats in real time, providing enhanced protection to both end users and financial institutions.

Behavioural Biometrics

Behavioural biometrics based on artificial intelligence is an advanced security method that adds an additional layer of user authentication by analyzing unique patterns of user behaviour with devices. Unlike traditional biometric methods based on physical attributes such as fingerprints or facial features, behavioural biometrics analyze softer aspects of user behaviour such as typing speed, mouse movement, and touchscreen behaviour. These behavioural traits are unique and can be reliably monitored to establish an active profile that mimics normal user behaviour. Through the application of AI algorithms to monitor these patterns in real time, behavioural biometrics can reliably verify user identity and detect suspicious behaviour or unauthorized access attempts. One of the main advantages of behavioural biometrics is its ability to detect anomalies in known behavioural patterns. Each user possesses a distinct way of interacting with their device that can be quantified in terms of traits like keystroke, mouse, and swipe patterns on touch devices. For instance, a user can type at a particular speed, employ certain finger movements on a touchscreen, or move the mouse in a particular way. AI algorithms can create a profile based on these traits and create a baseline of each user's normal activity. When a person uses the device, the system compares the real-time data against the established baseline in order to detect any abnormalities. If the system identifies a notable deviation from the user's established behaviour observable through metrics like reduced typing speed, variance in mouse movement, or uncommon touchscreen gestures, it can prompt an alert or demand extra authentication procedures. These procedures may involve the use of a one-time password (OTP) received on the user's mobile phone, or emailing the use of two-factor authentication (2FA) or conducting a biometric scan to add extra authentication of the user's identity. Such deviations can be modelled quantitatively using a distance-based measure such as Euclidean distance, which calculates the difference between the user's current behaviour and their established behavioural profile.

Euclidean distance is a mathematical term that is applied widely in fraud detection and behavioural biometrics to measure the difference or similarity between a user's real-time behaviour and their pre-learned normal behaviour pattern. In fraud detection different behavioural biometrics such as typing speed, mouse movement, and touchscreen interactions are monitored continuously to identify patterns specific to an individual. When a new interaction is made, the system compares it with the user's history of behaviour to determine whether the new input is in line with their normal behaviour or whether it exhibits extreme differences that may be fraud. Euclidean distance precisely measures these differences by calculating the straight-line distance between two points in an n-dimensional space where each dimension is a different behavioural feature (e.g., typing speed, between-key intervals, or pressure on a touchscreen). The formula for Euclidean distance is given as

$$d = \sqrt{(x1 - y1)2 + (x2 - y2)2 + \cdots + (xn - yn)2} \tag{8.10}$$

where

*x*1, *x*2, …, *xn* are the user's typical behavioural characteristics (e.g., typing speed, mouse speed, etc.)
*y*1, *y*2, …, *yn* denote the features found in the ongoing session.

Low Euclidean distance would imply conformity of the behaviour to the user's defined profile, while a high Euclidean distance would imply deviation of a high order, possibly indicative of fraud or unauthorized access. This methodology is extremely beneficial to implement in real-time fraud detection as it enables AI-fuelled systems to immediately identify anomalies, which triggers additional authentication processes (e.g., multi-factor authentication) or halts the suspected transactions. Leveraging Euclidean distance within artificial intelligence-powered fraud detection, banks and security solutions are able to gain increased precision, reduced false positives, and a dynamic security model that learns about users over time. Besides this, AI-based behavioural biometrics can utilize machine learning algorithms such as support vector machines (SVM) or DNNs to refine the user's behaviour model with time. As individuals use their devices, these models can learn and develop new behaviour patterns making fraud detection more effective. For instance, a person may alter their typing behaviour over time and the AI system can learn these alterations by updating the baseline model in real time. The formula for updating the machine learning model in such a scenario could be:

$$\theta_{new} = \theta_{old} + \alpha \cdot \nabla J(\theta) \tag{8.11}$$

where

θ _old denotes old model parameters,

α is the learning rate that shows how much the model adjusts with each new exchange,

$\nabla J(\theta)$ is the gradient of the cost function indicating how the model parameters should change to minimize errors.

By this ongoing process of learning, the AI system becomes increasingly effective at differentiating between genuine deviations (e.g., slight changes in user behaviour) and fraudulent activity (e.g., when an unauthorized individual tries to access the account).

Artificial intelligence-based behavioural biometrics provides a strong way of augmenting security through continuous measurement of user activity to ensure authenticity. By observing typing speed patterns, mouse activities, and touchscreen activities, AI is able to detect outliers from standard behaviour and trigger a response in accordance, such as requesting additional authentication procedures.

The utilization of distance-based metrics like Euclidean distance enables the system to measure the level of departure from the standard behaviour of the user and machine learning algorithms to adjust automatically to emerging patterns to enhance the level of accuracy in detecting fraud. The multi-layered solution delivers an effortless but very effective way of safeguarding users against unauthorized access while minimizing interference with the user experience.

Predictive Analytics

Artificial intelligence (AI)-driven predictive analytics has emerged as an indispensable tool for financial institutions to anticipate and pre-empt fraud threats before they inflict serious losses. Through examination of past records, AI models can detect patterns and trends that typically lead to fraud, allowing financial organizations to proactively respond. The use of predictive analytics in anti-fraud applications is centred around three functionalities: detecting high-risk customers and accounts, determining the likelihood of fraud prior to approval of transactions, and maximizing the use of fraud investigation resources through prioritization based on higher probabilities of fraud. Detection of high-risk customers and accounts is among the major uses of predictive analytics in the financial industry.

Through the application of machine learning algorithms, financial institutions are able to scan through enormous amounts of past transaction history and identify regular patterns that correspond with high-risk behaviour. Customers who regularly have frequent big transactions or tend to log on to their accounts from unfamiliar geolocation areas may, for example, be labelled as high risk. Predictive models may rate every customer or account with a risk score depending on account history, transaction anomaly frequency, and previous fraud incidents. The risk score can be calculated using formulas based on past data, like logistic regression or decision tree-based models to estimate the probability of fraud based on certain features. For example, the probability P of a client being high risk can be estimated as

$$P\left(High\,Risk\right) = \frac{1}{1 + e^{-(w1.x1 + w2.x2 + \ldots + wn.xn)}} \tag{8.12}$$

where $x1$, $x2$, ..., xn are customer behaviour-related features (e.g., number of transactions, login times, geolocation), and $w1$, $w2$, ..., wn are weights learned by the model from past experience. In case the risk probability is more than a given threshold, the system marks the customer as high risk, enabling the bank to track these accounts closely.

Determining the likelihood of fraud prior to approving transactions is another very important aspect of predictive analytics for real-time fraud detection. By examining data from past transactions, AI models are able to determine the likelihood that a new transaction is a fraud prior to approval. This is especially good at stopping fraud before any loss of money as it allows financial institutions to refuse suspicious transactions in real time. Financial fraud detection tools use predictive analytics to find unusual patterns of user behaviour that might be evidence of fraud. The main principle here is studying the history of transactions, geolocation, and how much they spend to see what's unusual. For example, if a client who has normally made low-level transactions within one area is now trying a big transaction somewhere else in the world, the system identifies it as an irregular behaviour pattern. Machine learning algorithms are used in AI-based fraud detection models to evaluate different parameters including transaction volume, device details, transaction time, and expenditure patterns to identify if the activity is valid or may be fraudulent. By comparing the new transaction against past data, the model determines a fraud probability score and responds accordingly. If the fraud probability is high, the system can automatically deny the transaction or ask for additional verification (for example biometric authentication or one-time passwords) [10]. If the probability is medium, it might alert the transaction for manual evaluation by fraud specialists. However, if the probability is low, the transaction happens normally. Sophisticated AI and deep learning algorithms can keep evolving to counter new patterns of fraud, minimizing false positives (legitimate transactions mistakenly identified as fraud) and false negatives (fraudulent transactions missed). Real-time fraud detection through this method enables financial institutions and online shopping platforms to increase security, reduce financial losses, and increase user confidence by allowing only authentic transactions to go through. A probabilistic model can be applied here where the likelihood of fraud *P*(Fraud) for a transaction is calculated as:

$$P(\text{Fraud}) = \text{Sigmoid}(w1 \cdot \text{Transaction amount} + w2 \cdot \text{Transaction location} + \cdots + wn \cdot \text{Transaction frequency}) \tag{8.13}$$

If *P*(Fraud) is more than a predetermined level, the system can indicate the transaction for additional verification or rejection. Implementing this predictive method, financial institutions can reduce false acceptance of fraudulent transactions thereby improving security.

Optimization of fraud investigation resources is a top issue for financial institutions since fraud investigations are time- and resource-consuming. Predictive analytics can be applied to rank the cases with the highest fraud potential and allow the investigators to focus their efforts on the most significant cases. For instance, if a predictive model predicts the likelihood of a specific transaction or customer being fraudulent, the system can push that case to be given an urgent priority. Alternatively, low-fraud potential cases could be deprioritized, resulting in saved effort and time. A priority score method can be utilized to assign a risk score for cases on the basis of transaction amount, user behaviour, and historical fraud. The score can be calculated by the addition of the various risk factors such as

$$\text{Fraud priority score} = \sum_{i=1}^{n} w_i \cdot \Delta x_i \tag{8.14}$$

where

x_i are the risk factors (e.g., transaction value, frequency of anomaly, age of account),

w_i are the weights given to every factor based on its importance in fraud prediction.

Fraud Priority Score higher cases are investigated first so that most likely fraud cases are cleared first and the low-scoring ones are processed afterwards or automatically cleared.

Predictive analytics enables financial institutions to be proactive and data driven in fraud prevention. By the use of AI models to identify high-risk customers and accounts, assess fraud risk before approving transactions, and automate fraud investigation resources, these institutions can significantly reduce fraud loss. Mathematical models such as decision trees, logistic regression, and probabilistic scoring systems are the foundation for the calculation of fraud risk scores and case prioritization for investigation. Predictive models not only heighten security levels but also make operations more effective, enhancing fraud prevention systems' efficiency and performance [11].

Automated Risk Management

AI is at the forefront of the automation of risk assessment and compliance screening, equipping financial institutions with the capabilities to detect and respond to regulatory breaking and combat financial crime more effectively. AI can scan transaction histories for violation of the law as one of the most important strengths. Through the analysis of massive datasets, AI systems identify patterns in behaviour that can signal non-compliance with the law such as money laundering, fraud, or tax evasion legislation. These types of transactions are flagged automatically by such systems, reducing the need for manual scrutiny and enabling compliance officers to concentrate on sophisticated cases.

AI can also ensure continuous compliance with changing regulatory standards through the ability to update its systems in line with new guidelines and regulations. In addition to tracking for cases of regulatory compliance breaches, artificial intelligence is also useful in detecting money laundering activities through advanced pattern detection. AI, by way of machine learning algorithms, can scan transaction streams to identify hidden relationships and anomalous patterns that could be indicative of illicit activities. These include the detection of structured deposits or withdrawals that are apparently contrived to evade detection, such as the fragmentation of large sums into small, less conspicuous transactions, a technique commonly used in money laundering schemes. AI systems can automatically flag these incidents for review, allowing for faster responses to potential threats and preventing large-scale money laundering schemes from escaping detection.

Artificial Intelligence greatly enhances traditional Know Your Customer (KYC) and Anti-Money Laundering (AML) processes [12], which are critical in ensuring that financial institutions do not unwittingly enable criminality. For KYC processes, AI allows for the verification of client identities by collecting data from various sources, including government records, credit bureaus, and social media platforms. This is a critical step in ensuring that financial institutions conduct thorough background checks on their clients, an important aspect of averting fraud and financial abuse. For AML compliance, AI technologies can trail the transaction data in real time, detect signs of suspicious transactions, and compare them to known historical patterns of fraud.

Automation of these processes not only increases operational efficiency but also increases the effectiveness of compliance activities, thus reducing the risk of human error and enabling faster responses to emerging threats. AI automates key aspects of risk assessment and compliance with regulations by clarifying the transaction data for violations, using pattern recall to find money laundering activities, and refining the KYC and AML processes. These AI activities notably improve the efficiency and veracity of compliance, allowing financial institutions to remain compliant with regulations and prevent the risks of financial crime.

Challenges and Ethical Considerations

While AI offers powerful tools for enhancing fraud detection, it also raises important challenges that must be carefully managed to protect individuals and organizations. Among these, ensuring robust data privacy and security is paramount to maintaining trust and complying with regulatory requirements in the financial sector.

Data Privacy and Security

Use of artificial intelligence (AI) in fraud detection has been found to have great potential for change, enabling financial institutions to quickly identify and counteract fraudulent transactions. But such technological advancement comes with extremely high challenges in data security and privacy. AI systems used in fraud detection procedures typically require access to large volumes of sensitive financial information, including transaction history, spending patterns, and customer identification details. Such delicate data is typically stored and processed on cloud-based systems, making it vulnerable to unauthorized access, hacking, or data breaches. It is therefore extremely critical for financial institutions to give high priority to the development of strong data protection mechanisms to safeguard such sensitive data from cybercriminals.

To comply with these regulations and minimize security risks, financial institutions must implement various data protection strategies. One of the most effective strategies is encryption, which renders sensitive data unavailable to unauthorized users. The process can be applied to the data stored (e.g., in databases) and data passed on (e.g., via the internet). For example, encryption techniques like the Advanced Encryption Standard (AES) [13], using a 128, 192, or 256-bit key size, provide high levels of data protection. The AES encryption algorithm can be expressed as:

$$C = E(K, P) \tag{8.15}$$

where

C is the ciphertext (encrypted data)
E is the ciphertext function
K is the encryption key
P is the plaintext (original sensitive data)

By using strong encryption methods, banks can ensure that if the information drops into the hands of unauthorized parties, it remains indecipherable and secure.

One of the most important security measures is multi-factor authentication (MFA) [14] for the users and administrators accessing AI-driven fraud detection platforms. MFA adds an extra layer of security because it requires users to authenticate themselves through various methods of verification such as a password, fingerprint scanner, or a one-time passcode sent on a mobile phone. This reduces the risk of unauthorized access, even if the login credentials get into the wrong hands [15].

Furthermore, access control mechanisms and user authentication systems need to be applied to guarantee that only the authorized personnel can access responsive to information. These mechanisms require the establishment of distinct roles and commitment within the company, where each worker is given access only to the information required for his or her respective job function.

This practice of segmenting data ensures that sensitive information is not inappropriately disclosed to those who do not have a required need for it.

Finally, institutions need to be careful about constant monitoring and auditing of their data protection processes. Frequent security audits and vulnerability scans of AI models need to be done to ensure systems are protected from emerging cyber threats. This ensures vulnerabilities that can be used by adversaries to attack are identified, and institutions take necessary action before the breach takes place. While AI-based fraud detection systems have the huge advantage of detecting fraudulent transactions in real time, they also carry huge risks to data privacy and security. Financial institutions need to have robust controls to guard sensitive customer data and comply with data protection regulations across the globe, such as the GDPR and CCPA. With a broad approach to data privacy and security, financial institutions can effectively balance the requirement for global fraud detection against the requirement to safeguard customer data. Methods like encryption, multi-factor authentication, and access control are necessary to safeguard financial information. With a broad approach to data privacy and security, financial institutions can effectively balance the requirement for global fraud detection against the requirement to safeguard customer data.

False Positives and Bias

The use of AI for fraud prevention has transformed the financial industry with greater accuracy and effectiveness in the detection of fraudulent transactions. Nevertheless, just like all cutting-edge technologies, AI models also come with challenges. One of the most serious challenges is the generation of false positives, wherein genuine transactions are mistakenly identified as fraudulent. This occurs when an AI model falsely identifies a transaction that closely follows patterns of proven fraud, although legitimate. An example is where a customer, who typically does small transactions, will have his transaction flagged as suspicious if he does a big transaction, yet the transaction is genuine. False positives can generate customer dissatisfaction, heightened operational expenditure, and even the disruption of genuine business transactions. Minimization of false positives is therefore an important component of any AI-powered fraud detection model. To false positives, a second top-priority issue is the bias of the AI models.

AI models are trained on past data, which, if biased or unrepresentative, can perpetuate discriminatory behaviour. For instance, if a model for detecting fraud is trained with past data significantly representative of groups, it will over-signal transactions by underrepresented groups as suspicious. Such bias could lead to discrimination against customer segments based on their race, gender, or socioeconomic status and result in unfair financial exclusion. To address this problem, caution must be applied in the training data to ensure the model is representative and unbiased across populations.

To avoid such problems from arising, ongoing model improvement is a must. Trained and deployed, the AI models must be updated on a regular basis with fresh high-quality data to keep up with changing fraud patterns and user behaviour. This can minimize the false positives and bias. Feedback loops can also be incorporated into the system to learn from previous errors. For instance, if a transaction was misclassified as being fraudulent, the system can re-examine this misclassification and refine its decision-making process so that it doesn't make the same type of error in the future. Another critical area is fairness auditing.

This is the process of auditing the AI model from time to time to certify that the model is making fair decisions and treating all the customers equally. Fairness auditing techniques such as appraising model outputs for different demographic groups can be applied to identify and correct any unfairness. Furthermore, the use of fairness measures such as Demographic Parity or Equalized Odds can quantify the level of fairness in the model's prediction. These measures can be used to ensure whether the model performance is consistent across different groups and whether some groups are being over-representatively classified as fraudsters. For example, Demographic Parity can be quantified as:

$$P(\text{flagged as fraud}|\text{group A}) = P(\text{flagged as fraud}|\text{group B}) \quad [8.16]$$

where

$P(\text{fraud}|\text{group A})$ is the probability that people belonging to group A are fraudulent,

$P(\text{flagged as fraud}|\text{group B})$ is the probability that group B is flagged as fraudulent.

If these probabilities are vastly different across groups, then the model is biased and needs to be tuned. In order to make the system fair, an open auditing process needs to be put in place that not only addresses reducing false positives but also eliminates any type of bias from the system. By doing this, financial institutions can make the overall accuracy of AI-based fraud detection systems higher while making sure all customers are treated equally and fairly.

Even while AI-based fraud detection software is providing strong safeguards to detect fraud it is not without problems like false positives and unfairness. There is a requirement for ongoing improvement of the model and for conducting fairness audits to rectify these problems. Banks can make their AI tools more efficient and equitable by conducting the periodic updating of training datasets, ongoing cycles of feedback, and the application of fairness measures.

Adversarial Attacks

As banks and other financial institutions become more dependent on AI-based fraud detection systems, fraudsters are getting more and more sophisticated in their efforts to bypass these systems. They keep evolving their methods to take advantage of weaknesses in the AI models and circumvent security controls. The most important issue in this regard is the phenomenon of adversarial attacks, where the attackers intentionally manipulate the input data or the model itself to mislead the system. These attacks come in a variety of forms, with data poisoning and model manipulation being two of the most significant. Data poisoning is the intentional process of fraudsters inserting false or malicious data into the training set of data employed for creating AI models. Through the insertion of misleading or misrepresentative data, they seek to have an impact on the way the AI system processes and classifies transactions. For example, a scammer might insert a group of fake but realistic-looking transactions into the training data to prevent the model from recognizing them, and so reduce the model's probability of detecting real fraudulent actions in the future. The model learns from this poisoned data, and its prediction errors and accuracy drop. An equation to express the effect of data poisoning on the performance of a model might be:

$$Accuracy_{pois} = \frac{True\ Positives_{poisoned}}{True\ Positives_{poisoned} + False\ Positives_{poisoned} + False\ Negatives_{poisoned} + True\ Negatives_{poisoned}} \quad [8.17]$$

where

- $True\ Positives_{poisoned}$ refers to the correctly identified fraudulent transactions after data poisoning.
- $False\ Positives_{poisoned}$ refers to the legitimate transactions mistakenly flagged as fraudulent.
- $False\ Negatives_{poisoned}$ refers to the fraudulent transactions that are not detected.
- $True\ Negatives_{poisoned}$ refers to the legitimate transactions correctly classified.

By manipulating the models, fraudsters can try and change the AI model's behaviour while it works. This would include changing a machine learning model's weights or the decision borders employed by the model to denote transactions as valid or suspicious. By changing the options, the fraudsters provide an easy environment for fraud not to be noticed, hence exposing the whole anti-fraud mechanism to risk.

To counter these sophisticated adversarial attacks, financial institutions must implement robust cybersecurity measures designed to protect AI models and their underlying infrastructure. The best way is making secure the data pipelines to prevent unauthorized access and implementing input validation techniques so that only legitimate data is given as input into the model. Implementing methods like differential privacy prevent models from learning personal or biased information during training. In addition, secure multi-party computation (SMPC) can also be utilized to share sensitive computations among several parties such that no party knows the entire set of data, further making the system robust against attacks.

Also, periodic model update is essential in order to fight changing fraud methods and protect against manipulative adversaries. As the fraudsters implement their tactics, the AI models also need to be retrained with new representative data continuously in order to accommodate emerging fraud trends. This repetitive process ensures the model is kept effective in recognizing new types of fraud, but also makes the model less susceptible to adversarial attacks. Furthermore, ongoing monitoring of model performance is able to indicate when a model is performing below expectations as a result of poisoning or manipulation and retraining should be undertaken forthwith.

Another important practice is adversarial training, where AI systems are intentionally subjected to adversarial examples during training. This helps the model learn to discern between genuine and fabricated data, better equipping it to resist adversarial attacks. An adequately trained AI system is more capable of identifying subtle manipulations or deceptive activities that might otherwise be overlooked.

As fraudsters persistently innovate and update their techniques to evade AI-based fraud prevention tools, the banks and other financial institutions should remain on high alert and active in their defences against adversarial attacks. Data poisoning and model tampering attacks can undermine the efficacy of AI models, and thus a robust cybersecurity process, frequent updating of models, and adversarial training are imperative to protect the integrity of the system. The use of such protective measures would ensure that AI systems in banks and financial organizations are resistant to new threats and offer a reliable and secure environment for fraud detection for their users.

Future of AI in Financial Fraud Detection

The promise of AI for preventing and detecting fraud is immense, especially with the persistent breakthroughs in various new technologies. One such technology is Explainable AI (XAI), whose focus is on rendering AI decision-making explainable. In the traditional AI models, especially the deep learning models, decision-making reasoning is normally hidden, and thus stakeholders, including regulatory authorities, cannot rely on or certify the system output. XAI attempts to counteract this by supplying reasons for decisions or predictions taken in intelligible language, including a rationale as to why a given decision or prediction was produced. This is especially vital in fraud detection since banks must be able to justify to regulators why specific transactions were identified as being fraudulent. Through supplying more transparent reasons for AI model decision-making, XAI allows for regulatory compliance helping financial institutions satisfy the increasingly growing legal and ethical requirements for using AI in high-risk environments such as fraud prevention.

Another exciting evolution in AI fraud detection is Federated Learning. This emerging method allows financial institutions or organizations to jointly train a common machine learning model without exchanging sensitive data. In the past, it would have been a matter of security and privacy for institutions to exchange financial data with each other. In federated learning, training of the model occurs locally at each institution based on data, and model updates instead of raw data are exchanged across the network. The collaborative approach fortifies the performance of fraud detection systems while upholding data security and privacy. When fraud trends vary across institutions, federated learning enables institutions to exchange intelligence and improve detection in a synchronized manner without undermining customer data.

With such innovations, Quantum Computing can revolutionize fraud detection systems with unprecedented processing power. Binary processing is used by traditional computers to solve problems, but quantum computers employ quantum bits (qubits) which can exist in multiple states simultaneously.

Quantum computers can thus perform complex calculations much more quickly than traditional computers. In fraud detection, quantum computing can scan enormous amounts of transactional data in seconds to identify complex fraud patterns that may not have been possible to detect. With its speed of processing, quantum computing can lead to easy fraud detection and shut down the window of opportunity for fraudsters to exploit loopholes.

Finally, AI-based Regulatory Compliance is one space where AI can actually make things easier for financial institutions. With mounting regulations on detecting fraud, financial institutions must make sure that their activities are in line with regulatory requirements. AI can make it easier to perform compliance checks by tracking transactions for potential fraud, continuously generating real-time reports, and flagging any activity against regulatory guidelines.

AI can also assist in making the process of reporting to regulatory bodies automated so that financial institutions remain up to date on their compliance tasks. Automating such processes not only removes the chance of human error but also frees up resources within financial institutions so that they can focus their time and efforts on more strategic anti-fraud programmes.

In summary, the future of artificial intelligence in fraud detection and prevention will be stronger and more effective with the application of explainable AI, federated learning, quantum computing, and AI-led regulatory compliance. These technologies will enable financial institutions to enhance transparency, enable collaboration with privacy-preserved access, use more computational power, and enable compliance automation. With the development of these technologies, they are set to provide a safer and more efficient platform for fraud detection and prevention in finance.

Conclusion

Artificial intelligence has transformed the detection of financial fraud by facilitating real-time monitoring, predictive analytics, and automated risk management, resulting in substantial advantages in fraud prevention and detection. These technologies enable financial institutions to detect anomalies in a timely fashion, predict possible threats, and better manage fraud-related risks. Nevertheless, these data privacy issues, AI model biases, and vulnerabilities to adversarial attacks need to be addressed. Banks must ensure that AI technologies are being implemented responsibly and ethically by transparent processes, stringent data protection, and constant monitoring to avoid biases and provide user privacy. Prevention of adversarial manipulation of AI models is also necessary to maintain system integrity. As AI keeps advancing, its potential to assist fraud prevention is enormous, but its deployment needs to be well managed to provide security, fairness, and regulatory compliance.

References

1. Faisal, N. A., Nahar, J., Sultana, N., & Mintoo, A. A. (2024). Fraud detection in banking leveraging AI to identify and prevent fraudulent activities in real-time. *Journal of Machine Learning, Data Engineering and Data Science*, *1*(01), 181–197.
2. Roy, S. S., Chopra, R., Lee, K. C., Spampinato, C., & Mohammadi-ivatlood, B. (2020). Random forest, gradient boosted machines and deep neural network for stock price forecasting: A comparative analysis on South Korean companies. *International Journal of Ad Hoc and Ubiquitous Computing*, *33*(1), 62–71.
3. Chahboun, S., & Maaroufi, M. (2021). Principal component analysis and machine learning approaches for photovoltaic power prediction: A comparative study. *Applied Sciences*, *11*(17), 7943.
4. Mori, R. (2021). Anomaly detection and cause analysis during landing approach using recurrent neural network. *Journal of Aerospace Information Systems*, *18*(10), 679–685.
5. Amato, E., Corradino, C., Torrisi, F., & Del Negro, C. (2023). A Deep convolutional neural network for detecting volcanic thermal anomalies from satellite images. *Remote Sensing*, *15*(15), 3718.
6. Lawi, A., Mesra, H., & Amir, S. (2022). Implementation of Long Short-Term Memory and Gated Recurrent Units on grouped time-series data to predict stock prices accurately. *Journal of Big Data*, *9*(1), 89.
7. Kuppala, L. (2024). *Custom-Named Entity Recognition Using Machine Learning* (Doctoral dissertation), California State University, Northridge.

8. Lin, A. K., et al. (2024). The AI revolution in financial services: Emerging methods for fraud detection and prevention. *Jurnal Galaksi*. https://doi.org/10.70103/galaksi.v1i1.5.
9. Talreja, V., et al. (2024). A survey of AI on financial fraud detection. *EPRA International Journal of Environmental Economics, Commerce and Educational Management*. https://doi.org/10.36713/epra16794.
10. Miao, Z., et al. (2024). Financial fraud detection and prevention. *Journal of Organizational and End User Computing*. https://doi.org/10.4018/joeuc.354411.
11. Nwankwo, E., et al. (2024). Reviewing the role of AI in fraud detection and prevention in financial services. *International Journal of Science and Research Archive*. https://doi.org/10.30574/ijsra.2024.11.1.0279.
12. Alexandre, C. R., & Balsa, J. (2023). Incorporating machine learning and a risk-based strategy in an anti-money laundering multiagent system. *Expert Systems with Applications*, *217*, 119500.
13. Kaur, J., Lamba, S., & Saini, P. (2021, March). Advanced encryption standard: Attacks and current research trends. In *2021 International conference on advance computing and innovative technologies in engineering (ICACITE)* (pp. 112–116). IEEE.
14. Mohammed, A. H. Y., Dziyauddin, R. A., & Latiff, L. A. (2023). Current multi-factor of authentication: Approaches, requirements, attacks and challenges. *International Journal of Advanced Computer Science and Applications*, *14*(1), 166–178.
15. Mohan, R., Boopathi, M., Ranjan, P., Na, M., Chaudhary, P. K., Chotrani, A. K., et al. (2024). AI in fraud detection: Evaluating the efficacy of artificial intelligence in preventing financial misconduct. *Journal of Electrical Systems*. https://doi.org/10.52783/jes.1508.

Chapter 9

The Role of AI in Combating Financial Fraud: A Case Study Perspective

Suhash Pokuri, Karri Hema Kumar, and
Abhichandra Polkampally

Introduction

The global financial world underwent radical changes during the last few decades because digital technology adoption reinforced internet-based financial transactions. Digital advancements created exceptional convenience for users of financial services, which allowed them to practice banking activities online and use mobile payment systems and operate across international financial boundaries. Financial systems face broad exposure to various types of fraud through digitization because it enables everything from basic credit card scams to grand theft operations led by international crime organizations [1].

Scope of Financial Fraud

Financial fraud encompasses a vast array of illicit activities. Other traditional forms of financial crime are credit card scams, theft of personal information, check kiting, faking money, and fraud in producing fake insurance. Modern

DOI: 10.4324/9781003606956-9

data phishing practices, digital platform growth, and fraudulent business sites, point-of-sale security breaches, and the malwares that extract consumer details have given birth to new kinds of cybercrime. Corporate data breaches such as these have an impact of millions on the finances of large organizations [2].

Limitations of Traditional Detection Methods

During past times, the financial establishments used a heuristic as well as rules-based system for uncovering fraudulent activities. However, previous methods of detection relied on domain experts to decide where to set up threshold boundaries or to specify hard-coded conditionals defining suspicious criteria such as all foreign activities that exceed $10,000. Such systems had ease of deployment, however, due to their simplicity, they suffered various crucial problems. Often, strict rules that otherwise would not cause errors on legitimate transactions would activate, leading to several false positive error occurrences and subsequent unnecessary investigations and customer dissatisfaction. The main foundation to build system rules was past fraud patterns, but fraudsters continuously made their methods deviate from these rules [3]. In periods of transaction growth and product expansion, technology was restricted by the lack of practical manual rule management of organizations. Rule-based systems detected little information, as they did nothing to detect the subtle or rare patterns that spread through different channels.

Rise of AI in Fraud Detection

Artificial intelligence was applied in fraud detection and an automatic detection system has been developed, which addresses some of the problems of the fraud prevention systems. Models which utilize artificial intelligence analyze large datasets (expressed in billions of transactions) to understand typical processes of the customer flow within and across different customer segments followed by the identification of out-of-bound segments. Learning to excel in classification, a machine learning (ML) classifier or deep learning architecture-based algorithm is as advanced as it can get compared to simple rule-based systems in terms of finding hidden complex relationships or nonlinear patterns [4]. Through the programming framework, it possesses the ability of continuous learning. Every time the system detects a true case of fraud and marked transaction, it increases its ability to predict future fraud cases, allowing for continual improvement of fraud prevention. In contrast, the operational flexibility of AI is the complete opposite of the very rigid rules which need human attention to activate. According to Zhang et al. [13], AI systems' capability to deal with unstructured database information such as text descriptions and chat logs or images enables complete fraud analysis.

Literature Review: AI in Fraud Detection

The evolution of fraud detection has been shaped by continuous advancements in technology and analytical methods. Early reliance on rule-based systems has gradually given way to sophisticated AI-driven approaches that can uncover hidden and evolving fraud patterns, transforming the way financial institutions safeguard against illicit activities.

Historical Developments and Paradigm Shifts

The first research in fraud detection prior to 2002 was through the use of statistical and rule-based expert systems. Finally at that time, the academic professionals collectively with their industry counterparts felt that these methods would aid in detecting the unusual patterns and deviations from recognized norms [5]. One of the causes for other flexible machine learning frameworks to receive focus is because of scalability 'challenges' and the suitability of these 'structures,' asking several researchers whether they are indeed suitable enough for scaling these 'framework' applications [3]. Now that they can recognize complex fraud patterns squashed by older methods, convolutional neural networks and recurrent neural networks can be used for effective planning of actions. As one of the most popular analytical tools used by researchers, generative adversarial networks are popular types of training systems used by researchers to recognize new types of fraud [2]. This has boosted supervised learning models which include support vector machines, decision trees, random forests, and many more. Also, we compared them with conventional traditional manual rules, wherein these machine learning techniques have shown better accuracy rates with minimum false alarms [6]. Today, researchers are very interested in the handling of complex financial records: analyzing high-dimensional data coupled with feature selection approach due to accelerating data rates, the growth of e-commerce, and mobile payment systems and digital banking.

Machine Learning and Deep Learning

Today, deep learning technology is distinguished because good computer systems and big data systems are available. Business transactions can be spotted with more complex hidden connections by models with multiple hidden layers. Thus, researchers adapted convolutional neural networks that are widely used in the regular image work for tables, dividing features into grid spaces. With transaction patterns, the popularity of Long Short-Term Memory (LSTM) networks has come from dealing with fraud patterns for multiple steps [6]. We used these neural networks to detect the time-based indicators that will lead to multi-step fraud activities. Despite this, the training data sample employed in AI systems is very small and it has inherent biases which are targeted at certain customer groups [4]. In order to effectively use AI systems for data detection, there are legal data privacy requirements, like European GDPR and California CCPA, that hamper the use of personal financial data. These days, generative models are effective in fighting fraud because they can generate synthetic scam transactions to improve model performance. Since fraudulent transactions are a small part of the entire banking activity, the new approach is shown to have much value in solving the class imbalance problem [6]. Today, researchers use graph neural networks to study the nature of bank account associations to locate money laundering networks and organization collaborations.

Key Challenges in AI-Driven Fraud Detection

A significant hurdle that still needs to be overcome through the use of AI systems to solve fraud issues is that they still have many problems. Finally, financial records break down into separate databases with sparse information on each transaction which makes them difficult to use as supervision in supervised machine learning; suppression relies quite a bit on excellent training material and financial records; however, without pointed exact details that quite often commerce financial records have, the data is not easy to manage financially [3]. Due to the difficulty in uncorrected data flow history and management standards, it is not possible for our evaluation model to detect trends or understand the derivation of the results. Normally, the number of fraudulent transactions out of total transactions handled by businesses does not exceed 1% or less. Specific approaches based on Synthetic Minority Over-sampling Technique (SMOTE) and cost-sensitive learning can be used to combat the reliance on most examples that occur when datasets possess significant minority imbalance but depend on most of the examples. When AI systems become popular, the adversaries of AI systems intentionally train selected examples in the hope of manipulating detection algorithms. As a result, the learning approximation between flawed detectors and attackers according to Khan [3] is obtained. Data privacy and protection laws cannot allow sharing of transactional data between institutions and within one institution, making it difficult to use data shared between institutions and making more thorough training (using more data than just observations of the training class) impossible. As the author notes in their work, businesses are required to receive customer permission to handle user data and keep the data within specified retention periods. Since the model processes all defense through history [4], it exaggerates risk for certain demographics. It is an unethical practice and could become a legal issue because the financial companies need to protect themselves from the discrimination, which they have been using against the customers.

Trends in Scholarly and Industry Research

Current research increasingly focuses on:

- Combined with multiple learners (weak learners combined such as a random forest and gradient boosting to increase detection accuracy).
- Techniques such as LIME (Local Interpretable Model-agnostic Explanations) or SHAP (SHapley Additive exPlanations) to explain how complex models decide [3].
- Streaming Analytics for Real Time: e.g. Apache Kafka or Spark Streaming for on-the-fly classification of thousands of transactions per second [6].
- Innovations like federated learning and secure multiparty computation, which leverage data-sharing constraints amongst financial institutions.

Comparative Analysis: AI vs. Traditional Fraud Detection

The performance metrics of AI-based fraud detection models display results superior to rule-based methods when measuring precision alongside recall and area under the curve (AUC). Traditional rule-based detection provides better interpretability and quicker deployment even though AI-based models have superior performance. Compliance teams typically select rules-based methods because they adopt clear rules while deep learning predictions require more complex justification efforts. Most practical deployments utilize information collected from multiple different sources. The main distinction emerges through comparing standard procedures with AI artificial intelligence models with static predefined traditional rules matched against AI dynamic learning capabilities. Traditional fraud detection approaches produce many incorrect positives while AI-based fraud detection systems achieve higher accuracy with lower incorrect predictions. The availability of adaptability varies between rule-based methods because they need manual updates while AI-based systems can adapt without human intervention through learning from new fraud patterns.

Summary of Literature Review

Admittedly, a lot of existing work proves that using AI-based fraud detection is more effective than older techniques. Current research is targeted toward enduring challenges such as interpretability, data privacy, and adversarial threats. Finally, the domain still holds many unanswered questions around ethics, governance, and robust deployment, but machine learning and deep learning have raised the bar for detection accuracy. The next sections show how the methodological and theoretical gains have to be studied in conjunction with the historical developments, the technical intricacies, the practical use cases, and the regulatory context to draw up one holistic path of progress.

Historical and Theoretical Contexts

The study of fraud detection has evolved through distinct phases, each building on previous insights to form a comprehensive understanding of combating financial crime. Early efforts focused on rule-based and statistical methods, setting the stage for today's sophisticated AI-driven approaches that adapt dynamically to emerging threats.

Early Fraud Detection Approaches

The history of fraud detection in financial institutions can be put back to the development of expert systems and statistical anomaly analysis in the second half of the twentieth century. Encoded domain knowledge consisted of the rules that were gleaned from the knowledgeable investigators, was utilized, and could scan transactional records for suspicious anomalies [5]. However, these systems were extremely rigid, and they were always forced to completely update their encoded rules when the fraud tactics had shifted. More data-driven methods were introduced due to the parallel advancement of statistical analysis. Logistic regression and time series analyses started to point to the correlation of fraud likelihood with properties of transactions. These statistical tools provided some flexibility, however, they could often struggle when there are more subtle and high-dimensional interactions present, typical of the modern fraud schemes.

Emergence of Machine Learning

In the 1990s, computational power grew and some machine learning algorithms such as decision trees and naïve Bayes started to play their role in fraud detection applications. These rules-based models performed better compared to the purely rule-based systems learnt using the labeled datasets and prognosis of the decision boundary-based metrics. This evolution represented a paradigmatic change as domain experts no longer had to explicitly encode each suspicion; the models were able to learn these automatically with the right amounts of historical examples; Support Vector Machines (SVMs) proved to be the state-of-the-art methods for handling class imbalance by maximizing the margin between classes, which helped improve robustness in classification tasks. SVMs were good with high-dimensional data, as well as very sensitive to outliers, as they had to be tuned extensively. Regardless, the change in their success proved that data-driven adaptive models would be the future that would shape fraud prevention rather than static expert rules.

Deep Learning Expansion

The 2010s brought the introduction of deep learning which meant that, besides lending a fair amount of strength to the field, currently neural networks can work on unstructured data such as text, images, and even voice recordings. For example, in fraud detection, RNNs such as LSTM capture recursive transactional data that identify patterns, such as repeating small charges in an effort to try to test new stolen credit cards. Autoencoders assess an anomaly based on reconstruction error while Generative Adversarial Network (GANs) provide a synthetic means of creating fraudulent examples to train more robust models.

Graph-Based Methods and Beyond

The second of these equally important developments is the use of graph theory to transactional data. Chains of money laundering across several accounts and groups coordinating various credit card attacks are typical and often carry with them the mark of fraud. Such relational dynamics can be modeled with Graph Neural Networks (GNNs) and hidden links or suspicious clusters in big transaction graphs become more evident. Additionally, the introduction of blockchain technologies gives new possibilities to the implementation of secure, distributed transaction ledgers upon which AI algorithms can work with limited data tampering [7]. While GNNs with blockchain-based data structures are a nascent field, they can be powerful, tamper-proof detection frameworks combined with advanced machine learning techniques and real-time analytics.

Evolution Toward Explainability

Gaining transparency in automated decision-making has led to an important theoretical advancement toward Explainable AI. Typically, these traditional machine learning and deep learning models log little insight into how they ever reached a hypothesis [3]. However, in the financial compliance requirements, legal liability, as well as public trust, enforces its interpretability. To explain a model, a number of techniques such as LIME, SHAP, and counterfactual explanations have been proposed. The next wave of the research may be bridging the gap of high model accuracy and comprehensibility in the long run.

Integration with Broader Theories

Recently, the area of AI-based fraud detection has been increasingly blurred with behavioral economics, criminology, and psychology theories. By examining instances of fraud, trust in financial transactions, or the actions of criminals impeded by fear of detection one can come to understand human factors and why individuals or groups commit fraud, forge documents, steal currency, or more [4]. For example, enhanced knowledge about social engineering tactics might help natural language processing (NLP)-based phishing attempt detection or malicious chat log detection.

Technical Details

Building effective AI-driven fraud detection systems requires integrating multiple technical components that handle data preprocessing, feature extraction, and model training. These components work together to transform raw transactional data into actionable insights capable of identifying fraudulent activities with high accuracy.

Core Technical Components

A suite of sophisticated techniques and tools comes in handy in building up AI-driven fraud detection.

1. Raw transactional data may have missing values, introduce noise, contain inconsistencies, etc. If I were to enumerate tools you use, I would have ETL (extract, transform, load) pipelines, validation scripts, and domain-specific transformation at the top of my mind. Feature engineering, e.g., time-lag between transactions, merchant categories, IP geolocation, generally leads to a significant increase in model performance in many scenarios.
2. Supervised Vs. Unsupervised: The make of choice depends on the existence of data. Supervised learning arises to supersede when there are plenty of labeled fraud turns, and unsupervised anomaly detection is beneficial for emergent deployments or whenever labels are expensive [7].
3. While classical ML models (such as decision trees, random forests, logistic regression) are robust enough, deep learning enables us to detect high-dimensional data and complex nonlinear relationships.
4. Real-time or streaming analytics architecture (e.g., Apache Spark, Kafka) allows for processing of suspicious transactions immediately. Conversely, batch processing may analyze the historical logs for patterns or retrain models on a scheduled basis [6]. In some advanced systems, Reinforcement Learning (RL) can be used. RL models learn an optimal policy by posing a tradeoff between false positives and missed fraud cases and get real-time feedback on the decisions.

Examples of AI Implementations

1. Note: A well-publicized case, PayPal uses deep learning to process the enormous transaction volumes of real-time anomalies detection to minimize the fraudulent transactions [2]. The low false positive rate of the system is maintained through continuous refinement of the system's model parameters based on confirmed fraud alerts.
2. Given the size of the company, the banking giant JPMorgan Chase applies AI-driven Anti-Money Laundering (AML) solutions with graph analysis and supervised learning for suspicious activity detection. The system builds transaction graphs to identify rings of associated accounts through layered transfers.
3. Mastercard: Introduced intercepting suspicious card present and card not present transactions using predictive analytics and deep neural networks as a part of Decision Intelligence. As a result, we are left with a sophisticated scoring system that provides merchants and issuers with the ability to quickly decide approval or disapproval for a transaction.

Advantages of AI-Based Fraud Detection

1. Many studies confirm that rule-based systems can be outflanked by AI models, especially deep learning, in terms of recall and precision, and are good at detection in many fraud cases where static thresholds are overlooking them [5].
2. Models can adapt fast to new fraud examples, making them adaptive with respect to new strategies.
3. Qualitative assessment of the risk to each transaction using appropriate computing infrastructures is possible in real time, thereby halting the fraud in milliseconds.
4. Less Manual Interventions: An ideal AI can significantly reduce the load of human investigators and therefore lower the operational costs [6].
5. AI solutions in cloud platforms and distributed computing frameworks can handle millions or even billions of daily transactions without performance degenerating [6].

Disadvantages and Limitations

1. Supervised Learning Hungry for Data: Free or expensive to get, much of the desirable data often lacks large volumes. Training the model with incorrect or biased labels can lead to corruption of the training [8].
2. Interpretability of Model: Many high performing AI models are often referred to as 'black box,' thus the inability to explain decisions to customers or even regulatory bodies [1].
3. Fraudsters may turn the tables on the AI models, supplying AI with special inputs (for example, slightly changing transaction attributes) to deceive the logic of detection.
4. Data sharing restrictions, privacy laws, and algorithmic bias constraints can hinder or require very complex compliance strategies.
5. Initial Investment: Creating a robust AI pipeline (data infrastructure, model training, specialized hardware) can be expensive making it difficult for smaller institutions or emerging markets [9].

Detailed Technical Example

- For instance, in a situation of credit card fraud detection within an international bank, consider the following.
- Real-Time Stream: Transaction data comes in as in real-time stream from ATMs, point of sale terminals, and online portals. Such records for user ID, card type, merchant details, geolocation, transaction amount, and timestamps.
- Additional Features: Also features extracted would-be time since last transaction, average spend patterns, location-based velocity checks, and risk scores of the merchant.
- Architecture of Model: A two-stage system is selected. An autoencoder first looks for anomalies by calculating the reconstruction error on each transaction that comes in. Once past a threshold, suspect transactions are classified by a supervised classifier (e.g., a random forest), and any above a threshold generate a suspicion via an alert.
- If an alert is found to be fraud by investigators, the data is labeled accordingly and periodically restrained with a supervised model to fine-tune future results [6].
- Outcome: The bank minimizes bad results on the validation data set and drastically reduces the false positive rate while minimizing false

negatives—the cases in which the scheme intercepts the fraud, but it is not a valid anomalous event [10].

Balancing Performance and Interpretability

And for financial institutions, this means that they may encounter a tightrope on which they will walk between maximum detection performance and transparent, explainable systems. There is an easier method to interpret such as decision trees or logistic regression, but in return the accuracy may be lower. Conversely, in the past, deep learning approaches usually achieved higher performance at a cost of interpretability [5] to address this tradeoff; the growing field of Explainable AI (XAI) tries to use methods such as SHAP values or LIME to explain the reasoning for each fraud prediction locally [3].

Future Technical Directions

- Models that adapt online to learn continuously, regardless of possible changing fraud patterns over time (i.e., without retraining from scratch).
- Pseudo Pseudo Partitioning: In federated learning frameworks, different financial entities can collaborate on training the models together without exchanging raw data [11].
- Combining NLP, GNNs, RL to solve multi-faceted fraud problems such as account takeover, and more elaborate layering in money laundering [4].
- Beyond Automatic Classification, AI Empowered Investigations: Such advanced AI could help human investigators with creating visual maps of linked entities or generate alerts of suspicious cross-border flow of funds [12].

Practical Applications

Credit card fraud detection represents one of the most critical and high-stakes applications of AI in finance. By analyzing vast amounts of transaction data in real time, AI systems help financial institutions quickly identify suspicious activities and prevent unauthorized transactions, protecting both customers and payment networks.

Credit Card Fraud Detection

AI solutions are tested by credit card fraud. The huge transaction volumes they handle mean that transaction decisions must be made within, many times, millisecond levels, and those are issued to banks and payment networks. Spend category, velocity checks, and location feature checks can be part of training models for the detection of suspicious transactions based on historicals [1]. One class of the classification methods that these banks use learns the normal spending of each cardholder and re-issues alerts only when a novel pattern differs enormously from established norms [13].

Anti-Money Laundering (AML)

The process of concealing the 'origin' or 'source' or 'source of funds' is money laundering. Human analysts had to comb through suspicious activity reports for traditional AML procedures. This process is automated and accelerated by AI, using a graph-based or network analysis technique to trace fine money flows [7]. For example, this is particularly important for large multinational banks processing thousands of daily cross-border transactions, so small incongruities between large amounts of data can be a clear sign that there is a laundering attempt.

Insurance Fraud Detection

Insurance fraud can occur in such different ways: it could be that of staged auto collisions, inflated claims on property damage, or false medical invoices. By cross-referencing claims data with external data sources such as hospital records, pre-existing claims history, social media, or even more explicitly, claims data with radar tech, machine learning helps insurers. Parsing textual claim descriptions may be used by NLP to detect inconsistent or contradictory accounts [14]. This results in suspect claims being flagged as a matter of deeper investigation, and therefore adjusters completing cases on legitimate cases dominated.

Stock Market Manipulation

It is illegal trading practices such as pump and dump schemes or insider trading that can affect market prices and disinter values in capital markets. Massive volumes of trading activity can thus be monitored by AI through time series analysis and sentiment analysis of the social media or financial news, and flags orchestrated hype or abnormal trading patterns [5]. Modern AI cannot be outfoxed in terms of granularity; it can evaluate short-interval trades, sometimes in microseconds, granularly. That makes it very helpful to regulatory bodies like the Securities and Exchange Commission (SEC).

Real-Time Transaction Monitoring

While some financial institutions only provide discrete financial products, many of them now have enterprise-wide real-time monitoring solutions that combine data from all business lines, such as savings accounts, investment platforms, credit lines, etc. Big data infrastructure such as Kafka or Spark Streaming extracts continuous data flows to build AI-based models that categorize transactions as a potential fraud in real time [4]. This integrative approach buttresses fraud-fighting tools since criminals typically exploit cross-product weaknesses in a single entity.

E-Commerce and Payment Gateways

CNP Transactions amount to high fraud risks for online marketplaces and payment gateways. Other than this, AI-driven systems can also validate IP geolocation consistency, device fingerprints, and user behavioral biometrics such as keystroke dynamics, and analyze the historical order pattern [10]. For example, if several orders are executed by the same user account coming from different countries in time, the system can prevent processing any transactions until identity verification.

Cryptocurrency Exchanges

The trend of cryptocurrency usage has seen new variety of fraud such as rug pulls, ICO scams, and exchange hacks. Anomaly detection based on AI is required to identify the suspicious wallet addresses or unusual transaction flow that is not behaving normally. Blockchain forensics and graph analytics are used to trace the movement of stolen or laundered crypto assets in several wallets [7].

Highlights of Practical Benefits

1. Automated AI screening frees investigators and biologists to focus on complex high-risk alerts.
2. Tailored models reduce false positives: Banks set a few legitimate transactions as false positives.
3. Cloud-based AI frameworks are scalable solutions which can be modified to settle the surges in the transaction volume without any major performance hits.
4. In sum, AI-based systems have carved out a transformative role in detecting fraud within financial services, and they have also shown their ability to demonstrate robust, AI-driven fraud controls to regulatory agencies to enhance an institution's standing. There are numerous case studies in which their success is evidenced, where advanced analytics saved losses from fraudulent schemes in the hundreds of millions, streamlined operations, and heightened trust in the customer.

Ethical and Regulatory Frameworks

Ethical and regulatory considerations are fundamental to the responsible deployment of AI technologies in fraud detection. Addressing issues of fairness, transparency, and accountability ensures that AI systems do not perpetuate bias or discrimination while complying with legal standards.

Ethical Considerations

The problem with fairness and ethics with respect to AI is that it always comes to the front. In particular, machine learning models they build using biased or unrepresentative training data can disadvantageously scrutinize or penalize particular demographic groups [4]. In the financial setting, such discrimination could appear in cases where a model is consistently determining that certain individuals from certain wards or socioeconomic backgrounds are 'bad,' because of historical prejudices in the dataset.

A second challenge is transparency; black box AI can indicate the determination of whether a transaction is suspicious, but is non-transparent to stakeholders, who may be customers, regulators, or the bank's own compliance officers who are trying to understand why a transaction was labeled suspicious. All of this can undermine the trust that underpins the system and affect the ability of wrongly flagged transactions to be appealed. However, these emerging fields like Explainable AI (XAI) aspire to explain deep learning for explaining the black box behavior, however, the complexity of the deep learning architectures is too forbidding for simple explanations [1].

Fraud detection is also closely related to privacy. A lot of personal data can be collected to rely on human involvement with AI-based monitoring, against the individual rights. Given the goal of preventing fraudulent activity, however, institutions must tread carefully to strike the proper balance between protecting the client and his rights to autonomy with respect to his client data and the right to be forgotten [6].

Regulatory Compliance

In many jurisdictions, laws have been passed to demand that organizations establish serious controls against fraud. At the same time, these entities have to abide by data privacy and consumer protection regulations, which in turn can be at odds with one's obligations. Key regulations include:

- The General Data Protection Regulation (GDPR) is enforced within the European Union, where strict guidelines exist pertaining to data use, compelling organizations to secure users' consent and grant customers the right to request removal of their data. Non-compliance can lead to hefty fines.
- Analogous to GDPR, U.S. state law gives Californian residents the right to know how they and their data are used, shared, and sold, as well as the right to demand their data's deletion.
- The principle of fairness and accuracy in credit reporting, and also related fraud detection mechanisms based on credit history.
- Financial institutions that provide international services have to follow AML directives (such as those of Financial Action Task Force (FATF) and their local equivalents), and this often requires a robust suspicious activity reporting supported by AI [8].

Balancing Security and Rights

Organizations have a dual imperative of strong fraud protection as well as data rights compliance. Better detection accuracy at the cost of privacy can be achieved over a very broad data collection. Meeting certain detection requirements, however, may depend on lean data usage, which might make detection ineffective. It takes well-defined data governance policies, audits for models, and sometimes uses techniques that protect privacy, such as encryption, differential privacy, or federated learning [12].

Compliance-Driven Model Management

One of the most important things in terms of being able to comply with regulatory standards is being able to track how an AI system is developed and updated. Model Risk Management (MRM) frameworks of institutions tend to make use of a Model Risk Management framework that tracks the version history, training data, performance metrics, and changes in model hyperparameters [3]. In addition, it also happens routinely to stress test the system in terms of its capability to withstand new fraud tactics or data shifts. Compliance is thus motivated to seek more responsible and monitored AI deployment.

Ethical AI Initiatives

This has led to the development of ethical AI guidelines and industry consortia due to growing awareness of algorithmic bias and privacy concerns. There are organizations like the Institute of Electrical and Electronics Engineers (IEEE), Partnership on AI, as well as the World Economic Forum that frame the issues of fairness, accountability, and transparency that help the organization. Compliance with these best practices would not only reduce legal risk for the financial institutions but also will build customer trust and the brand reputation [5].

Case Studies and Future Trends

Real-world case studies highlight the critical role AI plays in detecting and preventing financial fraud and data breaches. Examining these incidents provides valuable insights into the effectiveness of AI technologies and areas for future improvement.

Case Studies Examples

Case Study 1: Equifax Data Breach (2017)

The high-profile incident put over 140 million personal and financial data at risk [12]. It was primarily a breach, caused by poor system patching, but it showed the importance of badly needed advanced intrusion and anomaly detection systems. If AI had been sufficiently integrated back then, it could have prevented Equifax from becoming a victim by spotting unusual data exfiltration or access patterns. Within the aftermath, credit reporting agencies across the world started to heighten their prioritization of AI-based real-time breach detection to restore their reputation and regulatory damage.

Case Study 2: Capital One Data Breach (2019)

More importantly, the recent Capital One incident, where hackers took control of nearly 100 million customer credit card accounts and applications clearly highlighted the importance of configurations in the cloud [11]. While the attack involved a misconfiguration, rather than a specific flaw in fraud detection, what happened in the aftermath of the breach made it clear that combined AI-backed security that unites on premise and the cloud is pivotal. Now, the bank began to upgrade, bringing machine learning systems to serve as a tool for monitoring firewall logs, transaction behaviors, and other patterns of its own employees.

Case Study 3: PayPal's Fraud Detection Evolution

One often cited example of such leadership is PayPal that processes hundreds of millions of transactions per day with minimal friction to the customer. Using real-time machine learning models, features such as device fingerprinting, user behavior patterns, and dynamic velocity checks are employed in their system. Each fraudulent case could, once confirmed, instantly update the underlying model, and through an internal feedback loop ensure this. This iterative approach proves to be the most efficient fraud loss reduction method, and it maintains the detection accuracy year over year in a fast moving marketplace.

Case Study 4: JPMorgan Chase's AML Initiative

Graph analytics are used by JPMorgan Chase to detect suspicious groups of transactions that indicate that money laundering is taking place. The bank's AI system analyzes the node relationships (e.g., shared IS, IP, or transfer patterns) and connects seemingly unrelated transactions over national and business unit boundaries. These AI-driven works have been successful in early detection of laundering rings, and in some cases discovered cross-border complexities that rule-based approaches have not and therefore been successful.

Future Trends

Explainable AI (XAI)

The banks and the financial service providers are increasingly aware of the need for transparency as they undergo heavy regulatory scrutiny. With the goal of 'human-readable' justifications for flagged transactions, Explainable AI Frameworks are designed to provide justification for flagged transactions to bridge the gap between black box neural networks and compliance requirements [1]. The capabilities of techniques as LIME, SHAP, or counterfactual explanations to show which features contributed to fraud predictions are very useful in handling customer disputes and responding to legal questions.

Federated Learning

With federated learning, different institutions are able to work on a collaborative training of models without having to reveal sensitive or proprietary data. In particular, the raw data can be made private, for instance, a consortium of banks can learn fraud patterns from each other jointly, while only exchanging model updates, not raw data [11]. The advantage of this approach is especially relevant for fighting massive scale fraud networks operating in and across multiple financial institutions and geographies.

Blockchain and Distributed Ledger Technologies

Data storage via blockchain is tamper evident, such that records of transactions are irretrievably tamper proof. Combined with AI, it is possible to detect where funds move across many hops with high confidence in the integrity of the data [2]. Despite not being fully matured, coupling between the AI-based anomaly detection and blockchain-based immutability could certify how financial institutions manage and verify transaction records.

Multimodal Fusion

Textual data, images (for example, ID document scans), voice biometrics, as well as sensor data, from both mobile devices and other 'wearing' devices is increasingly being integrated into fraud detection systems. Specialized neural architectures can drive AI models that can process inputs simultaneously using multiple modalities (aka, multimodal inputs)—which might identify clever frauds that no single modal system can make out [13]. In identity verification and remote onboarding process, this approach is particularly useful.

Adversarial Robustness

The fact that fraudsters actively probe detection systems for weakness means that future solutions need to incorporate adversarial training (a technique where models are pestered with manipulated inputs aiming to mimic real attack) [4]. Such a process allows adaptation of detection engines by inoculating them against newly emerging evasion tactics.

Regulatory Tech (RegTech) Integration

Specialized RegTechs have ballooned second hand, and have since been worked by governments and financial authorities as they intensify oversight. Some make use of AI components to conduct compliance tasks, risk assessments, suspicious activity reporting, and so on. Shifting the focus from fraud, toward avoiding losses to theft, fraud promotion, and other types of risk, it will be increasingly difficult to differentiate fraud detection, compliance, and broader measures of operational risk management, where integrated AI solutions will save institutions significant costs and complexities [15].

Synergy of Human and AI Collaboration

Currently there is an emerging consensus that pure automation strategy is rarely the best strategy. Often, human–AI collaboration leads to better results, such as machines processing large-scale data and humans given contextual judgment. Augmenting human expertise seems to be the future of it (investigators tackle the highest impact, edge cases, and AI doing the routine triage).

Convergence of Thematic Directions

Combined, these trends indicate the future of strong, collaborative, and open AI-driven fraud detection systems. Financial institutions have a good chance to thrive in theoretically changing threat landscapes if they invest in data governance, continuous model improvement, and ethical frameworks.

Conclusion

The speed of digitization of financial services and its bringing unprecedented convenience and scale also gives opportunity to fraudsters to exploit systemic vulnerabilities fast. Like any other industry, there is no lack of interest in FinTech among the players participating in the adoption spree. Through this chapter, we have followed AI's development in fighting against financial fraud since it started with a literature review, the major breakthrough being methodological works ranging from simple statistical outliers to the sophisticated deep learning and graph-based models. We showed how the field gradually moved away from a brittle but rule-based approach that each arena forced upon us, to a more data-driven and self-adaptive approach guided by principles of adversarial learning and anomaly detection. We showed how the field gradually moved away from a brittle, rule-based approach imposed by each arena toward a more data-driven and self-adaptive paradigm, guided by principles of adversarial learning and anomaly detection. This shift has been enabled by advances in supervised and unsupervised machine learning, autoencoders, generative adversarial networks, and real-time streaming pipelines, which together help preserve both the high accuracy and scalability of modern detection mechanisms.

AI is demonstrated to be practical in various financial domains such as credit card fraud, anti-money laundering, insurance claims fraud, manipulation in the stock market, and cryptocurrency exchange fraud. However, these solutions have been successfully integrated into many institutions to reduce costs, improve the detection rates, and comply with very stringent regulatory standards. The chapter also discussed ethical and regulatory frameworks in relation to pressing concerns such as data privacy, algorithmic bias, and transparency. GDPR, CCPA, and AML directives are asking institutions to take care of the personal data in responsible ways, and providing justifications of automated decisions are some of the challenges that AI practitioners are solving through privacy-preserving computation, fair model training, and Explainable AI (XAI) research.

Analysts find on the one hand both the catastrophic risks from smarter-than-we-might-think cybercriminals and on the other hand the transformative potential of good old-fashioned robust AI. The advancement of AI threats requires parallel development of countermeasures by AI solutions. The chapter explores multiple important trends which will determine the development of AI-based fraud detection systems for future generations.

1. XAI enables organizations to achieve both model accuracy and the ability to interpret and audit their systems.
2. Many institutions can collaborate through Federated Learning which keeps their sensitive data protected from exposure.
3. Blockchain Integration: Leveraging tamper-resistant ledgers for improved transactional integrity.
4. Multimodal Data Fusion: Incorporating textual, visual, and behavioral inputs into holistic fraud detection architectures.
5. Advanced organizations develop strong protective measures to combat advanced fraud perpetrators who deliberately try to breach machine learning systems.
6. The strength of automation should be combined with human expertise to solve complex cases because expert investigators provide essential context and judgment.

AI-driven fraud detection offers organizations a fundamental transformation of financial institution capabilities to detect, analyze, and fight fraudulent activity within their digital operation environments. A global financial system that functions better through resilience and inclusivity can be created when stakeholders combine their expertise across historical knowledge and advanced theory with the best technical processes and ethical principles and innovative methods for the future. Continuous research activities together with collaboration between stakeholders will enhance AI's capabilities for financial transaction protection, thus creating a future system that detects and suppresses fraud quickly to protect consumers along with businesses in every part of the world.

References

1. Aggarwal, C. C. (2017). *Outlier analysis*. Springer.
2. Alpaydin, E. (2020). *Introduction to machine learning* (4th ed.). MIT Press.
3. Barredo Arrieta, A., et al. (2020). Explainable Artificial Intelligence (XAI): Concepts, taxonomies, opportunities and challenges toward responsible AI. *Information Fusion*, 58, 82–115. https://doi.org/10.1016/j.inffus.2019.12.012.
4. Binns, R. (2018). *Fairness in machine learning: Lessons from political philosophy*. Proceedings of the 2018 Conference on Fairness, Accountability, and Transparency, 149–159.
5. Bolton, R. J., & Hand, D. J. (2002). Statistical fraud detection: A review. *Statistical Science*, 17(3), 235–255.
6. Bose, R., et al. (2022). Big data analytics in financial fraud detection: A framework for real-time transaction monitoring. *Journal of Big Data Research*, 9(2), 111–125.
7. Casino, F., et al. (2019). Cómo NFT pueden volverse valiosos. *Information Processing & Management*, 57(2), 102–134.
8. Chauhan, H., & Patel, N. (2021). Financial models suffer from attacks produced by adversaries. *Cybersecurity Journal*, 5(1), 19–34.
9. Goodfellow, I., et al. (2016). *Deep learning*. MIT Press.
10. Xu, Y., et al. (2021). Behavioral biometrics for online banking security. *Cybersecurity & Privacy*, 2(3), 21–37.
11. Yan, G., & Han, S. (2022). Graph neural networks in fraud detection: Theory and practice. *IEEE Transactions on Neural Networks and Learning Systems*, 35(4), 987–1003.
12. Yang, Q., et al. (2019). Fed: Federated learning for financial fraud. *Journal of Machine Learning Research*, 20(1), 1–22.
13. Zhang, Y., et al. (2021). Natural language processing in financial fraud detection: A systematic review. *Expert Systems with Applications*, 176, 114–125.
14. Zhang, Z., et al. (2022). A meta-learning approach to anti-fraud in the financial networks which evolve quickly. *IEEE Transactions on Computational Social Systems*, 9(2), 513–528.
15. Zhou, R., et al. (2019). A review of adjudications and problems: AI-driven insurance fraud detection. *Insurance Technology Review*, 10(4), 22–35.

Chapter 10

Neural Correlates of Impulse Buying: A Causal AI Approach to Consumer Preference Reversals

Ruslana Raemgulova, Alexander Didenko, and Egor Volkov

Introduction

Impulsive buying behavior is a complex economic and psychological phenomenon that exerts a profound impact on the global economy and consumer culture. Research indicates that 40–80% of all retail purchases worldwide occur impulsively, without prior planning, amounting to hundreds of billions of dollars annually. This phenomenon spans across all categories of goods and services, from everyday necessities to luxury items, and manifests in both traditional retail environments and e-commerce platforms. Impulsive buying behavior holds interdisciplinary significance, standing at the intersection of marketing, consumer psychology, behavioral economics, and neuroscience. The relevance of research in this field is driven not only by economic consequences for businesses and consumers but also by the socio-psychological aspects of contemporary consumer culture.

The economic impact of impulsive purchases is substantial, given their share in total global retail sales. According to Liao and Wang [20],

DOI: 10.4324/9781003606956-10

approximately 62% of supermarket sales are classified as impulse purchases. The economic magnitude of impulsive purchases varies by retail format and product category. In e-commerce, impulsive purchases constitute about 40% of all transactions, exceeding $175 billion annually in the United States alone. The 2021 statistics show that 86% of consumers make an unplanned or impulsive purchase online.

Research in neuroscience and cognitive psychology reveals the fundamental mechanisms underlying impulsive buying behavior. Studies published in the *Journal of Neuroscience, Psychology, and Economics* demonstrate that impulsive purchases are associated with activation of the limbic system, responsible for emotional responses, and a simultaneous decrease in activity in the prefrontal cortex, which governs planning and self-control. Functional magnetic resonance imaging (fMRI) shows that impulsive purchase decisions trigger dopamine release—a neurotransmitter linked to pleasure and reward, explaining the emotional satisfaction consumers experience from spontaneous acquisitions [1].

Psychological research identifies four primary types of impulsive buying behavior: pure impulse buying (entirely unplanned and emotionally driven), reminder impulsive buying (triggered by sudden recall of a product need), promotional or suggestion impulse buying (response to a perceived deal), and planned impulsive behavior (intended unplanned purchases under certain conditions). Individual personality traits also significantly influence susceptibility to impulsive buying. A large-scale study in *Frontiers in Psychology* found that high scores in impulsivity, novelty-seeking, and materialism correlate positively with a tendency toward spontaneous purchases. Surprisingly, even individuals with high financial literacy exhibit impulsive buying tendencies, emphasizing the strong emotional influences that bypass rational considerations. According to a Gallup Institute survey, approximately 84% of consumers admit to making emotion-driven purchases, often perceiving them as a form of mood regulation.

This study investigates the causal relationship between neural responses to pricing errors, as measured by error-related negativity (ERN), and subsequent consumer purchase intentions. In particular, we aimed at identifying the EEG correlates of impulsive buying behavior depending on [1] external situational online-purchase stimuli and trait impulsivity among non-clinical adults. For that, a two-stage Virtual Shopping Task was designed in an ERN paradigm to capture consumer preferences and preference reversals, realistically imitating consumer experience on an online marketplace of apparel. We further implemented a classic Stroop Task with EEG recording and the

Barratt Impulsivity Scale (BIS-11) to evaluate individual impulsivity based on self-reported, neural, and behavioral data.

Initially, we review the neural, cognitive, and behavioral aspects of impulsivity as predispositions toward impulsive buying behavior, with regard to the implemented experimental paradigm. Further, impulsive buying as preference reversal is discussed, followed by a conceptual framework, hypotheses, and experimental procedure. Research implements Causal AI data analysis, as well as Explainable AI (XAI) methodology: Shapley Additive Explanations (SHAP) and CatBoost. The results have revealed a strong causal relationship between the neural activity detected in the frontocentral and orbitofrontal brain areas associated with processing the discrepancy between subjective product price estimates and market prices and subsequent willingness to buy. With the use of advanced machine learning methodologies, these factors and nudge-based stimuli for purchase were found significant in predicting preference reversals, or impulsive buying.

Method

To better understand the mechanisms underlying impulsive purchase decisions, we first examine the neural and cognitive underpinnings of impulsivity as a stable individual trait. This section outlines the theoretical and empirical foundations that guided the selection of experimental tasks and assessments used in our study.

Neural and Cognitive Aspects of Impulsivity

Trait impulsivity, as defined by Moeller, is a predisposition toward rapid, unplanned reactions to internal or external stimuli, often without regard to the potential negative consequences. It is considered a relatively stable personality trait, although it might change in the course of life. Impulsivity manifests behaviorally in an inability to inhibit prepotent responses (impulsive actions), a preference for immediate gratification, and a tendency to discount larger delayed rewards (impulsive choice) [1]. Impulsive individuals tend to prefer risks; they are more novelty and sensation seeking. These tendencies significantly affect consumer decision-making that may lead to unfavorable outcomes for individuals and households [2]. Thus, impulsivity has been studied in both clinical patients (e.g., ADHD, depression, anxiety, substance abuse disorder) and healthy subjects across psychology, genetics, neurophysiology, and other fields [3].

Trait impulsivity is often discussed as a multidimensional construct comprising urgency, perseverance, non-planning, motor impulsiveness, premeditation, and attention components. Psychometric assessments were developed to evaluate these domains, including Eysenck's Impulsiveness Inventory (IVE), UPPS-P Impulsive Behavior Scale, and Barratt Impulsivity Scale (BIS-11). BIS-11 has demonstrated internal consistency, test-retest reliability, and construct validity. However, its factor structure was repeatedly found inconsistent [4].

In addition, individuals who score high on impulsivity questionnaires may not necessarily demonstrate more impulsive responses in behavioral tasks, as self-reported traits do not always align with observed impulsiveness. Findings from Go/NoGo experiments indicate that self-reported impulsive subjects can produce similar results to the low-impulsive group in reaction times and rate of accuracy. However, they demonstrate increased P300 latencies, probably because high-impulsive people have to exert greater effort for response inhibition [3]. In addition, components of trait impulsivity measured by BIS-11 correlate differentially with proactive/reactive cognitive control during Stroop Task. For instance, motor impulsiveness is associated with greater proactive control, while attention impulsivity is linked to reduced proactive control [5]. These findings highlight the necessity for multimodal assessment, combining self-report, behavioral tasks, and neurophysiological measures—to capture the complexity of the trait with higher accuracy.

The neural mechanisms underlying impulsivity and impulse control include the prefrontal cortex and several corticostriatal loops. Deficient frontal-cortical regulation over impulses and limbic drive has been linked to attention deficits and impaired behavioral and cognitive inhibition in the subclinical population [6]. Bari and Robbins proposed a model of interaction between executive functions emphasizing the importance of attention and inhibition for impulsivity. They conceptualize impulsive behaviors as emerging in response to external or internal changes in environment as impulsion, or "urge to perform a specific task" an individual fails to control [4]. This and other dual-process models view impulsivity as a "joint outcome of reflective and associative underlying neural processes" [7].

According to Nigg's taxonomy, inhibition encompasses interference control identified as the "ability to maintain response performance during the presentation of competing stimuli." Interference control underlies goal-oriented behavior as it is involved in monitoring and evaluation of conflict induced by task-irrelevant information from the environment that disrupts the ongoing activity. It might be specifically important for consumer decision-making, as consumers are exposed to advertisement, visual cues, affective messages, and other external triggers for purchase.

Stroop task is a traditional method to evoke interference and examine one's ability for interference control. The stimuli in a classic color-word Stroop task include four names of colors (e.g., 'blue', 'red', 'green', 'yellow') printed in different font colors, matching or mismatching the semantic meaning of the word. Subjects are required to press a button corresponding to the font color of a stimulus. Interference occurs when the meaning of the word is incongruent to its color (e.g., 'red' written in green). Hoedlmayr et al. studied the neural correlates of interference control during a classic Stroop task. They identified that N400 potentials registered in the frontocentral brain area are more negative in response to incongruent than congruent stimuli 400–500 ms post-stimulus onset. Research suggests that this so-called N400 Stroop effect is a manifestation of higher cognitive cost of the response in the presence of interference [8]. Hence, interference control, as measured by N400 in Stroop task, can be a potential predictor of trait impulsivity as far as inhibition of automatic response is concerned for both. In other words, greater cognitive effort exerted to suppress interference can be tracked in N400 potentials to incongruent words and can be informative of individual impulsivity. Capturing behavioral, self-report, and EEG measurements will allow us to examine the effect of interference control ability and trait impulsivity—as relatively constant factors—on impulsive buying behavior.

The transition from making unique impulsive decisions to developing impulsivity as a trait can occur via reinforcement learning, as "behavior-specific impulsive responses are learned" with experience [9]. Built on that, impulsive people differ a lot in sensitivity to some stimuli but not others. Therefore, impulse buying in response to certain situational factors can be habituated in consumer experience forming into impulsive tendencies. These patterns might predetermine an impulse control disorder, as the initial decision to engage in an activity is voluntary, but eventually an individual experiences urges to repeat the behavior that cannot be controlled .

In this study, we used BIS-11 and Stroop task to analyze how trait impulsivity and interference control contribute to impulsive choice in Virtual Shopping Task. For that, we rely on self-report, behavioral, and EEG assessments. Therefore, domains of interest are interference control, motor impulsiveness, attention impulsiveness, and non-planning.

Impulsive Purchase

Stern suggested classifying impulse purchase into: (a) pure impulse buying, (b) reminder impulse buying, (c) suggestion impulse buying, and (d) planned impulse buying [5]. Further, Wolman described impulsive behavior as non-planned, but momentarily triggered by external stimulus exposure. Building on Wolman, Rook and Rook & Hoch stated that buying urge is spontaneous, "powerful and persistent." It is typically combined with psychological conflict and struggle, and reduced cognitive evaluation of decision consequences. An impulse "is more likely to involve grabbing a product than choosing one" [10].

The majority of experimental studies on impulsive buying behavior, online and offline, adopted these views, applying the Stimulus–Organism–Response paradigm. For example, Zheng proposed that scarcity claims elicit the impulsive urge to buy a product. In contrast, website properties, such as excessive choice variety and the quantity of information, can undermine the impulse due to deeper cognitive evaluations of options. Aesthetic website qualities are specifically crucial for markets of apparel due to the associated hedonic property [9]. Besides, Jeffrey and Hodge demonstrated an increase in "impulse item" purchases with an increase in the amount of money spent by individuals prior to the suggestion [11].

Many of these inquiries discuss cognitive and affective aspects of impulsive consumption in a broad variety of contexts, thus contributing to the understanding of impulsive buying behavior. However, they often heavily rely on post-purchase self-report/interviews in assessing impulsiveness, not considering its neurophysiological substructure [3, 9]. In contrast, neuromarketing EEG-based studies do discuss consumers' neural response to marketing triggers in e-commerce, but outside of impulsive buying discourse [12]. Therefore, there is a significant conceptual gap between these two research directions. This is in part because impulsive buying is spontaneous and differs from person to person, which makes it problematic to simulate in a controlled environment.

Conceptualizing impulsive buying as preference reversal under the effect of external stimulus allows for such experimentation with the inclusion of EEG correlates associated with willingness to buy. In line with Kahneman and Tversky's theory of reference-dependent preferences, Hoch and Loewenstein examined in an experiment the causal relation between time-inconsistent preferences and consumer self-control. Accordingly, impulsive buying can manifest into "momentary and involuntary departure from the consumer's dominant preferences." So, consumers experience "being driven to buy something against their own better judgment," as they demonstrate inconsistent preferences with self-control as a mediating factor [13].

From the neuroeconomics perspective, preferences and preference reversals can be observed under the Reinforcement Learning Theory of ERP (event-related potential). Schaefer et al. investigated the relationship between Price Prediction Error, its size and valence (Large/Small; Underpriced/Overpriced), and consumer preferences. Significantly overpricing a product elicits a positive surprise during the market price exposure, causing increased willingness to buy. Further, neural response to market prices is analyzed via peak-to-peak difference of N200 and P300 EEG components in feedback-locked ERP waveforms in frontocentral electrodes, revealing a direct link between positive price surprise and choosing to buy. We implemented this paradigm into our design to capture consumer preferences dependent on the expected against the actual price difference. It is in line with other research showing that the N200 EEG component accurately represents consumer preferences [13]. So, observation of EEG activity in response to price provides additional validity of preference measurement beyond self-report.

Notably, preferences and judgments are highly sensitive to social influence, often leading individuals to adjust their behaviors to align with societal norms or the opinions of others [10]. This suggests that preference reversals can be significantly shaped by external social cues. Furthermore, numerous studies have identified neural signatures of such preference changes, demonstrating how social influence modulates value representations in brain regions like the ventral striatum and orbitofrontal cortex [7]. As a minor study objective, we observed the mutual impact of self-reported propensity to conform and susceptibility to marketing cues representing social judgment about a product, on preference reversals [14].

Just like impulsivity being multidimensional, online impulsive buying behavior varies significantly in underlying motivation and demonstrated decision-making, personal traits, sensitivity to triggers, habits, and circumstances. Grabbing a product might be viewed as pure impulsive purchase and correlate more with prepotent motor responsiveness to sensory input, while making a purchase off-budget aligns with delay discounting. Therefore, Stop Signal/GoNoGo, Stroop/Flanker Task, and discounting tasks—widely used to measure the underlying neural mechanisms of impulsive action—could all be applicable paradigms to evaluate the neurophysiological component of impulsive buying, depending on specific conceptualization and context of choice [15]. Therefore, because the different types of impulse purchases highlighted in the classification of Stern reflect the various degrees of cognitive evaluation preceding impulsive buying, we focused on the suggestion impulse purchase [16]. Thus, we attempted to analyze impulsive buying more from a cognitive rather than affective standpoint, as the change of consumer preferences due to external nudge-stimuli for purchase with mediating personality factors—trait impulsivity and trait conformity [17].

Conceptualization and Hypotheses

We conceptualize impulsive buying as preference reversal [16]. The primary determinants of an impulsive purchase are (1) trait impulsivity as a relatively stable consumer characteristic and (2) situational or "nudge" stimuli for impulsive purchase (e.g., scarcity cue, social cue, and price discount) [18]. Their relative contributions to preference reversal are analyzed in our Virtual Shopping Task, which simulates an online-purchase environment to reveal and potentially reverse consumer preferences (see Appendix A.1 for an overview).

In Stage 1, we aim to (1) verify the association between price prediction error and consumer preferences by replicating Schaefer et al.'s experiment design, and (2) reveal participants' initial preferences for subsequent reversal [19]. In each trial, participants view a product image and state their subjective value for the product. Next, the product's market price is shown, and they decide whether to purchase it by clicking "buy" or "not buy." By observing the difference between subjective value and market price, we evaluate price prediction error. Error-related negativity (ERN) locked to market price exposure indicates price surprise induced by this prediction error [19]. Thus, buying decisions serve as a behavioral operationalization of consumer preferences.

We expect that the magnitude of price discrepancy will inform the degree of price surprise. Particularly, a positive difference between subjective value and market price will elicit a positive price surprise (observed in increased ERN) and lead to stronger willingness to buy. Conversely, a negative discrepancy will cause negative surprise and subsequent unwillingness to buy. Therefore, we hypothesize that ERN in response to product market price will predict participants' "buy" or "not buy" choice. We also propose that ERN locked to market price exposure is mediated by trait impulsivity,

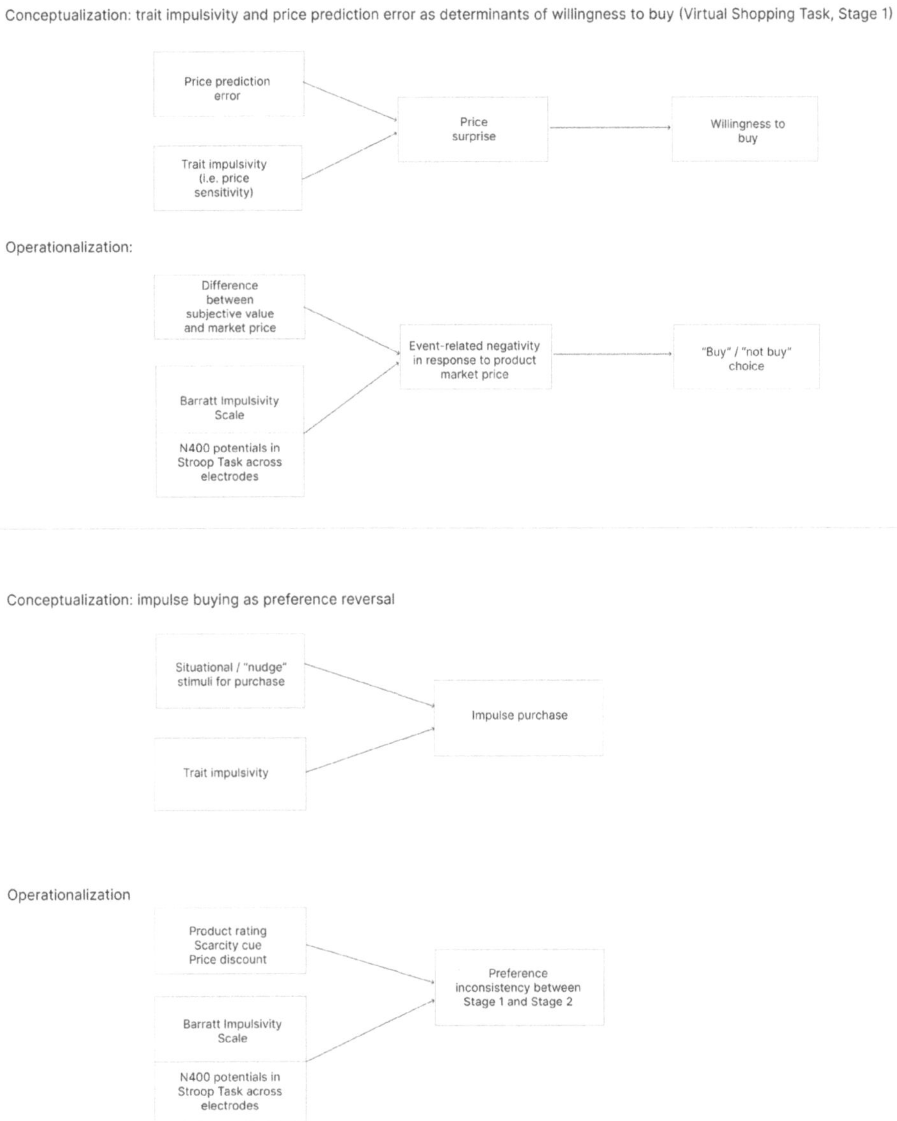

Appendix A.1 Conceptual scheme of research.

given that impulsive individuals often exhibit heightened price sensitivity [20].

In Stage 2 of the Virtual Shopping Task, we aim to reverse consumer preferences, with preference reversal operationalized as choice inconsistency between Stage 1 and Stage 2. Participants are presented with pairs of previously preferred and non-preferred items, along with additional information such as social ratings, scarcity cues, and a discount on the non-preferred product [21]. This alternative framing is designed to nudge participants toward reversing their initial choices. As the departure from initially stated preferences under these stimuli aligns with impulse buying in prior research, we regard such reversals as indicative of an impulse purchase.

Trait impulsivity is operationalized on both behavioral and neural levels via (a) self-reported impulsivity measured by the Barratt Impulsivity Scale (BIS-11) and (b) the N400 potentials in the Stroop Task across electrodes, which reflect cognitive control over impulses [21]. We hypothesize that trait impulsivity will mediate the neural response to product market prices in Stage 1, as impulsive individuals are more sensitive to price discrepancies. In turn, these neural responses are expected to predict subsequent buying decisions. We also anticipate that participants with higher impulsivity will be more prone to preference reversals in Stage 2. By hypothesis, BIS-11 scores and N400 peak amplitudes over frontocentral sites in the Stroop Task will predict participants' choices in Stage 2 of the Virtual Shopping Task.

Although several studies have experimentally examined impulse buying behavior [22], to our knowledge, this is the first to analyze EEG activity associated with impulse buying in a controlled yet naturalistic setting designed to simulate marketplace shopping.

The conceptual framework of the study is provided in Appendix A.1 for reference.

Procedure

Prior to experiment participation, all subjects (n = 87) familiarized themselves with the instructions and gave informed consent. A subsequent survey ensured normal or corrected to normal vision, absence of psychiatric or neurological conditions, control for sex, handedness, and age. The Russian version of BIS-11 followed immediately after, as well as the test for conformity. Next, subjects underwent 1-minute closed-eye condition before the two-stage Virtual Shopping Task. The latter was constructed via PsychoPy software. Throughout the experiment, subjects' EEG activity was being recorded by brain–computer interfaces Neuroplay-6C (electrodes: Fp1, Fp2, O1, O2, T3, T4; sampling rate: 250 Hz). N = 33 out of 87 participants underwent classic Stroop Task with EEG recording, on a separate day.

Virtual Shopping Task

Experiment stimuli comprised two sets (for men and women) of 20 images depicting selected items from Asos.com marketplace online-catalog in the category 't-shirts and vests,' balanced to represent the supply in this category across brands, color, absence/presence of print, fabric, and price. Participants could choose one of the two task versions: with men/unisex or women/unisex apparel. The experiment interface, from a participant's viewpoint, is given in Appendix A.2.

Stage 1 included 20 trials. Product image was projected for 5 s. Then, we asked the participants to offer their price in a typing window. If the subjective price differed from the market price by more than 20%, this response was removed from the data: subjects received a message "too low"/"too high," and the next image was demonstrated. It was done to prevent strategic underpricing. Conversely, when the expected price met the threshold, the market price was demonstrated for 5 s. Here, we captured the price differences and neural price prediction error processing (1500 ms feedback-locked ERN in response to market price). Lastly, subjects viewed the product image again with choice options: "buy"/"not buy." This decision allowed us to divide the items into two groups according to preferences, for each participant. All steps were separated by a 5 s fixation cross [23].

Stage 2 comprised 10 trials. Randomly joint pairs of products from the preferred and non-preferred categories derived at the previous stage were demonstrated with stimuli: 'items available,' 'item rating,' 'old price,' and 'new price' for the non-preferred item. Participants responded to a question 'Which one to buy?' by clicking either option.

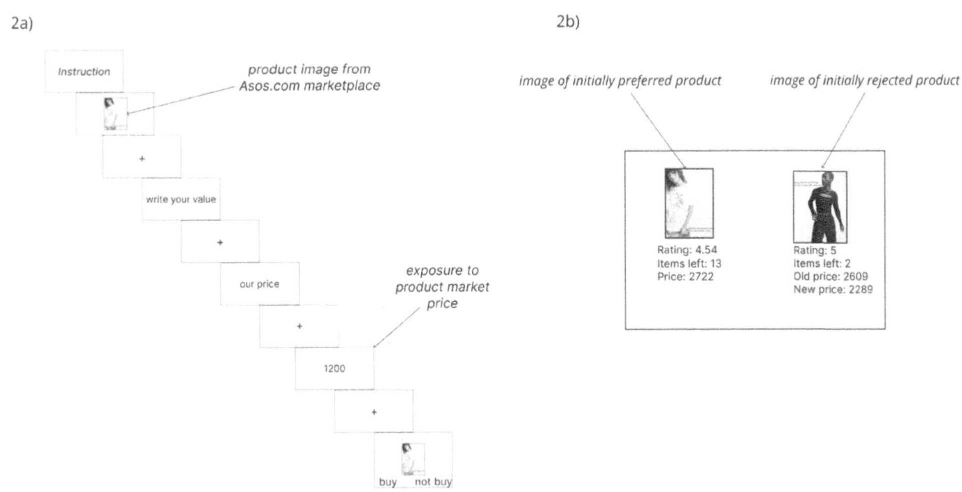

Appendix A.2 Virtual shopping task: participants' view: 2a) Stage 1 trial example; 2b) Stage 2 trial example. The left product in 2b was previously preferred, and the right product was previously rejected (images retrieved from https://www.asos.com).

Analysis

EEG Data Preprocessing

For EEG activity in Virtual Shopping Task, Independent Component Analysis (ICA) was implemented to preprocess the data and remove the artifacts (i.e., facial muscle movement, blinking). FRN was derived by computing the peak-to-peak difference between N200 and P300 feedback-locked components over frontocentral and orbitofrontal brain areas in response to market price demonstration in time windows (–200; 1500) ms for each stimulus exposure [24].

ML Workflow Overview

Explainable AI (XAI) methodology was applied for the prediction of consumer preferences and preference reversals, while Causal AI techniques were used for additional investigation of causal relationships between variables of interest. The usage of six-electrode brain–computer interfaces (BCI) enables the collection of data from more experiment participants due to its ease of operation. Embracing our "simple device–complex algorithm" paradigm, we leverage the benefits of these accessible devices by integrating them with advanced causal and explainable AI techniques to compensate for their inherent hardware limitations. Nevertheless, it can impose significant limitations to the quality of data and the coverage of brain sites of interest. While classic data analysis methods might not produce meaningful results, explainable and causal AI analysis techniques provide an opportunity to partially mitigate equipment drawbacks.

Particularly, gradient boosting algorithms like CatBoost are capable of capturing 'nonlinear interaction effects between the features' when dealing with categorical data [25]. In addition, Shapley Additive Explanations (SHAP) is a useful methodology to evaluate the importance of each feature on the model output as well as assess its predictive efficacy.

Initially, we employed CatBoost and SHAP to analyze the data for Stage 1, focusing on predicting consumer preferences. Using similar methods, we examined the effect of the trait impulsivity features on the neural response to market prices, specifically for products preferred in Stage 1. Subsequently, we aggregated data from Stage 1 and Stage 2 of the Virtual Shopping Task, Stroop Task, and BIS-11 to predict preference reversal using CatBoost, SHAP, and Causal AI. All models were developed using the CatBoost algorithm, selected for its ability to handle categorical data and capture nonlinear interactions between features effectively. Model efficacy was evaluated using performance metrics such as accuracy, F1-score, precision, recall, and ROC AUC score. To ensure robustness, we applied Shapley Additive Explanations (SHAP) to interpret feature importance and assess the contribution of individual features to the model's predictions [26].

Initially we used two target features: buying decision in Stage 1 ('bnb') and choice between the previously preferred and non-preferred options in Stage 2 ('which_buy'). These analyses can be viewed as binary classification problems. Both outcomes were highly imbalanced. In 'bnb' outcome, the choice 'buy' had a low proportion in the dataset compared to 'not buy' that was approached through upsampling the minority class to match the majority class (see implications further). For the 'which_buy' outcome, the minority class was the choice of the previously preferred item, meaning, participants rarely selected products in correspondence with their initial preference statement. The class imbalance was resolved by using SMOTE (Synthetic Minority Over-sampling Technique) and subsequent application of class weights.

Machine Learning Models

In this section, we present a series of machine learning models developed to predict key behavioral outcomes based on neural, cognitive, and situational features. Each model addresses a specific research question derived from our conceptual framework and utilizes explainable and causal AI techniques for interpretation and validation.

Model 1: Predicting Preference Reversal

This model aimed to predict consumer decisions to buy or not buy a product. Key features included Sub_Val, representing the subjective value participants assigned to a product, Price, reflecting the market price, and ERN activity across frontocentral and orbitofrontal electrodes (Fp1, Fp2, O1, O2, T3, T4). The data was divided into training and testing subsets in a 0.65:0.35 ratio from 87 subjects [27]. Model development was done in two entries: first, the model was trained on the original, imbalanced data, and then on the upsampled, balanced data. In both cases, the model was trained with a learning rate of 0.25 over 100 iterations. The selected features included

'Sub_Val', which represents the subjective value that participants assigned to a product upon initial exposure to its image. 'Price' feature represents the market price of the product. ERN activity in response to price presentation was represented through features corresponding to specific electrodes (Fp1, Fp2, O1, O2, T3, T4). To assess the classification ability of both we calculated the performance metrics: accuracy, F1-score, precision, recall, and ROC AUC score.

Accordingly, prior to upsampling, the model reached accuracy and precision of 0.88, F1-score of 0.80, Recall of 0.74, and ROC AUC score of 0.93. These metrics indicate relatively strong predictive performance, but potential bias toward the majority class ('not buy'). We further upsampled the minority class to balance the dataset and recalculated the metrics, arriving at the following results: accuracy of 0.69, F1-score of 0.275, precision of 0.63, recall of 0.176, and ROC AUC score of 0.68. A decrease in accuracy and recall indicate the remaining difficulty in identifying minority cases. However, precision of 0.63 might suggest that the model predicts 'buy' instances more accurately post upsampling [28].

We applied SHAP (Shapley Additive Explanations) to evaluate feature importance in predicting impulse buying behavior. The SHAP summary plot (Figure 10.1) provides a list of the ten most influential features affecting the model's output. Each point represents an individual prediction, with feature impact (SHAP value) shown on the x-axis. A positive SHAP value indicates an increased likelihood of preference reversal, whereas a negative value suggests the opposite. The color gradient reflects feature values, with high values represented in darker and low values in lighter gray.

The SHAP analysis highlights several key factors influencing impulse buying behavior. Scarcity cues (lefton1) emerged as the most impactful feature, demonstrating that the lower availability of previously preferred products significantly increases the likelihood of preference reversal. The positive SHAP values indicate that scarcity cues serve as strong nudges for impulse purchases. Attentional impulsivity (BISattn) also plays a crucial role, with individuals scoring higher on this trait being more susceptible to preference reversals, although its effect varies depending on other interacting factors [29].

Additionally, the order in which products were presented (index1, index2) influenced decision-making, with earlier exposures having a stronger effect on preference formation. Cognitive control, as measured through the Stroop Task (stroopImp), was another determinant, where lower cognitive control correlated with a greater likelihood of preference reversal. Price sensitivity

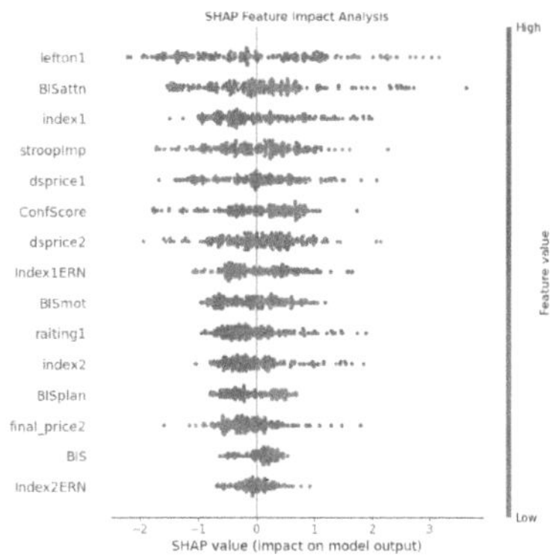

Figure 10.1 SHAP feature impact analysis for preference reversal (impulsive behavior) prediction model.

(dsprice1, dsprice2, final_price2) further played a significant role, with pricing variables, particularly discounts, affecting buying behavior.

Social conformity (ConfScore) also contributed to decision reversals, with higher conformity scores making individuals more sensitive to external cues and reinforcing the role of social influence in purchasing decisions [30].

Overall, the findings indicate that the attentional impulsivity subscore from the Barratt Impulsivity Scale (BISattn) played a more significant role in predicting preference reversals than the total BIS score (BIS). This suggests that attentional impulsivity, rather than general impulsivity, is a stronger determinant of impulse buying behavior. Interestingly, this result is partially supported by the Stroop Task interference score (stroopImp), which primarily measures motor control and cognitive interference resolution. The fact that attentional impulsivity was a stronger predictor than motor impulsivity challenges the traditional view that impulse buying is purely a failure of response inhibition [4, 21]. Instead, it suggests that impulsive buying may stem from an inability to sustain attention on initial preferences, making individuals more susceptible to external marketing nudges such as scarcity cues, social influence, and pricing strategies [31].

This aligns with the idea that impulsive purchases are not necessarily the result of a lack of inhibitory control over actions but rather a cognitive vulnerability to distraction by new information. The weaker role of more complex cognitive control measures further reinforces this hypothesis, indicating that higher-order executive functions may be less involved in spontaneous buying decisions than previously assumed.

Lastly, neural responses to price prediction errors, specifically error-related negativity (Index1ERN, Index2ERN) signals measured during pricing exposure, were found to be significant predictors of preference reversals. This finding supports the neuroeconomic perspective that links violations of price expectations to consumer decision-making, suggesting that individuals who exhibit stronger ERN responses to pricing discrepancies are more prone to shift their preferences when exposed to external marketing nudges [13]. However, the exact role of ERN in this process remains unclear. While it is traditionally associated with performance monitoring and error detection, its contribution to impulse buying may reflect a heightened sensitivity to price discrepancies rather than a direct causal driver of preference reversals [32].

To clarify the relationship between ERN components and impulse buying behavior, we evaluated a series of additional models, which are discussed in the following sections. While the SHAP analysis confirms that Index1ERN and Index2ERN contribute to predicting preference reversals, their precise functional role remains unclear. Whether ERN primarily reflects sensitivity to price prediction errors, heightened cognitive conflict, or an underlying predisposition to shift preferences impulsively requires further examination [33].

Model 2: Predicting ERN with Self-Reported and Neurobiological Impulsivity

This model was developed to predict Index1ERN, representing the error-related negativity (ERN) response to product prices that were later chosen in Stage 1 of decision-making. The key predictors included ERN responses to congruent (cong_peak) and incongruent (incong_peak) stimuli recorded in the Stroop Task across six electrodes, as well as trait impulsivity scores from the BIS-11 scale.

The model achieved an R^2 score of 0.4978, indicating a moderate predictive performance. Figure 10.2 depicts SHAP feature impact analysis for ERN prediction. Feature importance analysis revealed that cong_peak_Fp1_x was the most influential feature, followed by cong_peak_O2_x and cong_peak_O1_x, suggesting that ERN responses to congruent stimuli over frontopolar (Fp1) and orbitofrontal (O2, O1) electrodes play a key role. Incong_peak_T3_x and cong_peak_T4_x were also among the top contributors, emphasizing that both congruent and incongruent responses influence the model predictions [33].

Regarding trait impulsivity, BISattn (attentional impulsivity) had a stronger impact than the overall BIS score (BIS), confirming that attentional impulsivity may be more relevant to ERN modulation than general impulsiveness.

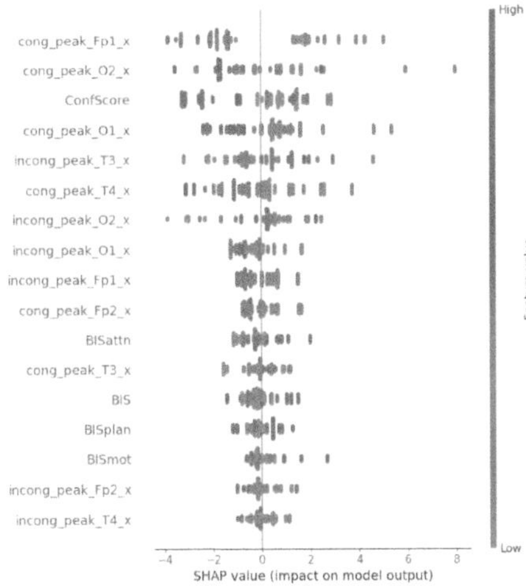

Figure 10.2 SHAP feature impact analysis for ERN prediction model.

BISmot (motor impulsivity) and BISplan (non-planning impulsivity) contributed minimally, suggesting that higher-order cognitive control mechanisms play a less significant role in shaping ERN responses in this context [34].

This analysis highlights the interplay between cognitive control, neural conflict monitoring, and impulsivity in response to pricing stimuli. However, to better understand the precise role of ERN in consumer decision-making, we further explored alternative model specifications in subsequent sections.

Models 3 and 4: Estimating Quasi-Uplift to Examine the Causal Impact of ERN on Preference Reversal

To investigate the causal effect of error-related negativity (ERN) on consumer choice behavior, we employed an uplift modeling approach. Specifically, we categorized participants into treatment and control groups based on their ERN responses during Stage 1 of the Virtual Shopping Task. The treatment group consisted of individuals exhibiting high ERN responses (above the median), while the control group included those with lower ERN values. Two separate machine learning models were trained using CatBoost classifiers to predict purchase decisions within each group. The uplift effect was then estimated as the difference in predicted probabilities of choosing a product between the treatment and control groups. The average treatment effect (ATE) was calculated as the mean difference in predicted probabilities, yielding an uplift effect of 0.1646 (Figure 10.3).

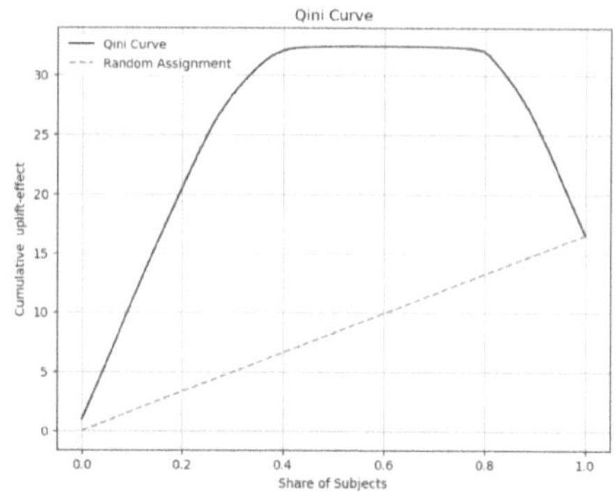

Figure 10.3 Qini curve for uplift analysis of ERN effect on preference reversal.

To check whether the uplift effect is statistically significant, we ran a one-sample t-test to see if the observed uplift differs from zero. The test gave a t-statistic of 2.6427 and a p-value of 0.0096. Since the p-value is below 0.05, we can confidently say that ERN does affect preference reversals. In other words, people with stronger ERN responses are more likely to change their choice when exposed to marketing nudges.

The Qini curve further illustrates the impact of ERN on purchase behavior. The solid curve represents the cumulative uplift effect, demonstrating the increase in expected purchases when prioritizing individuals with higher predicted uplift scores [35]. The bottom line represents a baseline scenario where treatment assignment is random. The fact that the Qini curve rises above the random assignment line indicates that ERN-driven interventions improve prediction accuracy and enhance targeting efficiency. The steep initial rise of the curve suggests that a subset of participants is particularly responsive to price prediction errors, implying that ERN might serve as a useful neurometric predictor for identifying individuals most susceptible to marketing nudges.

Analysis of Modeling Results and Causal Inference

The present study aimed to investigate the neural nature of impulse buying as preference reversal. Thus we examined the comparative impact of (a) trait impulsivity features (N400 evoked potential in Stroop Task and BIS-11), (b) neural response to price prediction error during primary product price exposure, and (c) situational stimuli for purchase in the form of scarcity cue, social cue, and discounted price—on preference inconsistencies. Our findings provide several insights into the causal relations between these factors, while outlining important methodological implications for future studies of impulsive consumer behavior.

In Stage 1 of the Virtual Shopping Task, we replicated the experimental design suggested by Schaefer et al. to examine the impact of neural processing of price prediction error on the willingness to buy [39]. Using gradient boosting algorithms and Shapley Additive Explanations (SHAP), our analysis indicated that error-related negativity (ERN) recorded over frontocentral and orbitofrontal brain areas during product price demonstration—as well as subjective valuation—play a significant role in predicting consumer preferences. These findings align with Schaefer et al.'s observations regarding price surprise and willingness to buy.

Further, we implemented nudge-stimuli to evoke preference reversal, that is, a behavioral operationalization of an impulse purchase. Firstly, we observed a clear instance of preference change, as behavioral results suggest. Accordingly, 499 out of a total of 546 choices at this step were made in favor of the previously rejected item, indicating preference reversal. This change often occurred even when discounted prices for rejected items significantly exceeded the unchanged prices of the previously preferred options. Following the "simple device–complex algorithms" paradigm, we applied categorical boosting to predict preference reversal using behavioral and neural features. We observed that Stroop interference and trait impulsivity modulate the neural reaction to prices of the preferred products, suggesting that the neural response to error directly influences behavior in the impulsive task. The explainable AI analysis revealed that scarcity cues, operationalized as lefton1, are significant for evoking preference reversals, as indicated by their high feature importance. BISattn (attentional impulsivity) was the second most important feature for predicting preference inconsistency, while BISmot (motor impulsivity) had a moderate impact, and BISplan (non-planning impulsivity) demonstrated a minor contribution. Additionally, strooplmp, representing ERN in the Stroop Task, also contributed significantly to predicting the which_buy outcomes [37]. Pricing-related features, such as dsprice1 and dsprice2 (initial market prices), showed that higher initial prices reduced purchase likelihood, while finalprice2 (discounted price) increased the likelihood of preference reversal. On a conceptual level, these findings suggest that scarcity cues, attentional impulsivity, interference control, and the interplay between initial and discounted prices may collectively play key roles in driving impulse buying, operationalized as preference reversal [38].

To clarify the relationship between ERN components and impulse buying behavior, we evaluated several additional models. SHAP analysis confirmed that both Index1ERN and Index2ERN contribute to predicting preference reversals, though their exact functional roles remain ambiguous. In Model 2, designed to predict Index1ERN using Stroop Task ERN responses (congruent and incongruent peaks across six electrodes) and BIS-11 trait impulsivity scores, we achieved an R^2 of 0.4978. Feature importance indicated that ERN responses at Fp1, O2, and O1, along with those at T3 and T4, were influential, while attentional impulsivity (BISattn) was a stronger predictor than overall impulsiveness. In Models 3 and 4, we employed an uplift modeling approach by categorizing participants into high and low ERN response groups and training separate CatBoost classifiers to predict purchase decisions. The estimated average treatment effect was 0.1646, and a one-sample t-test ($t = 2.6427$, $p = 0.0096$) confirmed that stronger ERN responses causally affect preference reversal. The Qini curve further demonstrated that individuals with higher predicted uplift scores are more likely to change their preferences when exposed to marketing nudges, underscoring ERN's potential as a neurometric predictor in impulse buying behavior.

Discussion

The present study is an exploratory entry into the laboratory investigation of online impulse buying behavior. Through our analysis, we have demonstrated that impulsive purchases account for a significant portion of global retail transactions [39], influencing profitability, marketing strategies, and consumer decision-making. The economic scope of impulsive buying extends across both traditional and digital, including mobile, commerce. The psychological, neuroanatomical, and neuroendocrinological mechanisms underpinning impulsive buying, including the role of the limbic system, dopamine release, and individual personality traits [4, 5, 7], underscore the complexity of this behavior and its deep-rooted connection to emotional regulation and decision-making biases. Recognizing that these mechanisms do not operate uniformly across consumers, impulse buying varies significantly between individuals with respect to the efficiency of external stimuli and internal driving factors determined by prior experiences. This variability, coupled with its spontaneous and unpredictable nature, renders impulsive buying particularly challenging to analyze in a controlled environment.

The proposed conceptualization of impulsive buying as preference reversal mediated by marketing cues, neural responses to price prediction errors, and individual personal characteristics serves as an entry point for developing an experimental procedure to study the neurophysiology of impulsive consumption. Our current research contributes to advances in neuromarketing, supporting evidence for the utility of neural measures, such as error-related negativity (ERN) and the N200—in capturing consumer preferences and responses to pricing stimuli.

From a methodological perspective, our study employs a two-model uplift modeling approach using CatBoost in the form of a T-learner to analyze consumer decision-making. The T-learner framework, which consists of separate models for the treated and control groups, allows for a structured estimation of treatment effects while accounting for differences between these groups. This approach effectively captures nonlinear interactions and provides more granular insights than simpler uplift techniques. However, it assumes that the available sample size is sufficient to train two separate models without overfitting, a challenge in cases of limited data [40].

A key challenge in studying impulsive buying behavior is disentangling the impact of price expectations and subjective value, as they are often conflated in operational measures such as "subjective price." The inability to clearly separate preference reversal based on price stimuli versus visual appeal introduces ambiguity: does a consumer reject a product due to its high price, or do they simply not want it at any price? One possible refinement would be to allow consumers to assign a subjective price of zero to certain items, clarifying their valuation process. Experimental adjustments, such as altering the sequence of stimuli presentation or adding non-price preference trials (e.g., a simple "like/dislike" task) could provide more reliable measures.

Existing research suggests that event-related potentials like N200 are particularly relevant for tracking preference formation [13], whereas price surprise is more indicative of shifts in valuation. Building on our established two-stage Virtual Shopping Task, the updated design will introduce changes in four aspects. One update is including additional within-stage trials where apparel items (e.g., T-shirts) are presented repeatedly with and without pricing information. These trials will allow us to record neural markers (e.g., N200) that specifically track pure visual preference. Control for income and other demographic variables will parse out external influences on subjective valuation and impulsivity. The next phase of research will further investigate how combined neural markers (ERN and N200) relate to both price-driven and visually affected consumer decisions. Additionally, future research should examine the influence of social conformity tendencies—measured neurally—on impulsive buying conceptualized as preference reversal [41].

Analysis methodology could include several causal AI techniques, from uplift modeling to graph causal AI, to establish the causal impact of neural responses on both initial purchase intentions and subsequent preference reversals, while testing hypotheses derived from behavioral economics and neuroeconomics. While the T-learner is a strong baseline for estimating causal effects in marketing and consumer behavior, it does not fully address issues related to hidden confounders, selection bias, or the causal mechanisms driving decision-making. More advanced causal AI techniques such as Propensity Score Matching (PSM), Double Machine Learning (DML), and Structural Causal Models (SCM) provide additional layers of interpretability and robustness. These methods enable researchers to refine causal inferences by explicitly modeling treatment assignment probabilities, correcting for observed and unobserved confounders, and simulating counterfactual scenarios.

A comparison of these approaches highlights the increasing levels of complexity and data requirements:

- T-Learner: Requires a reasonably large dataset with clear separation between treated and control groups. It is effective when the groups exhibit distinct characteristics but it is sensitive to sample size limitations.
- Propensity Score Matching (PSM): Balances treatment and control groups by matching units with similar likelihoods of receiving treatment. While it reduces selection bias, it demands a sufficiently large dataset to ensure adequate matching without data loss.
- Double Machine Learning (DML): Uses machine learning models to estimate treatment effects while adjusting for confounders. It is computationally demanding and requires strong assumptions about the validity of adjustment variables.
- Structural Causal Models (SCM): Explicitly define causal relationships using a directed acyclic graph (DAG), allowing for counterfactual inference. SCM requires domain expertise to construct meaningful causal structures and a rich dataset with well-defined variables.

Another critical factor is income as a control variable and its role in balancing experimental conditions within experimental setups [42]. Income may influence both price sensitivity and impulsivity, making it necessary to incorporate as a covariate. Preference reversals observed in our study may be attributed to habituation effects, lack of incentives, or the nature of the product set (e.g., disliking most available options). The absence of a preference commitment may explain why consumers easily switch their choices in Stage 2. Potential refinements include balancing designs to account for such effects or making these considerations explicit in participant instructions.

The conceptual framing of impulsive buying can be further enriched by theories from Kahneman and Tversky, particularly regarding changes in decision-making mode and alternative framing of similar options [43, 44]. To better capture these effects, experimental designs should aim to equate decision conditions across trials, ensuring that the manipulation of stimulus attributes directly aligns with theoretical constructs such as reference dependence and the endowment effect. Similarly, Hoch and Loewenstein's framework on time-inconsistent preferences and preference reversals provides a strong foundation for understanding impulsive buying as a departure from initial preferences against consumers' better judgment [16]. Future research should operationalize these constructs by incorporating measures of urgency, emotional arousal, and decision regret to assess the full spectrum of impulsive decision-making.

While our current approach relies on T-learner uplift modeling, future research should explore more advanced causal inference methods that go beyond estimating treatment effects to identifying underlying causal mechanisms. The integration of classical causal discovery with innovative architectural paradigms such as neurochaos learning and graph-based representations offers a promising direction for causal modeling in consumer behavior research [27]. These methods can enhance our ability to understand how impulsivity emerges and manifests in decision-making, particularly in the context of neurophysiological data. Techniques such as Structural Causal Models (SCM) and Double Machine Learning (DML) offer frameworks for integrating multi-level causal reasoning into neurophysiological modeling. These approaches enable a more systematic evaluation of the causal pathways driving impulsivity, moving beyond associative patterns to explicit intervention-based reasoning [45].

The next step in this research will further build upon our "simple device–complex algorithm" approach, leveraging accessible EEG technology combined with advanced causal inference techniques. The paradigm of "simple device–complex algorithms" holds significant promise for the field of neuromarketing, as it enables the collection of neurophysiological data using accessible, cost-effective tools like Muse and Emotiv while leveraging sophisticated analytical techniques to uncover complex patterns in the neurophysiological mechanisms of consumer behavior. Consumer-grade EEG devices [46], due to their relative affordability and ease of use, democratize the process of data collection in neuroscience research; Sabio et al. (2024) highlight that "the primary research-related use for consumer-grade EEG devices is for BCI systems," which indicates a shift whereby simpler, more accessible devices can accommodate advanced analytical methodologies, thereby facilitating robustness in research outcomes [37]. By integrating neuronal recordings with state-of-the-art causal discovery algorithms, we can refine the understanding of reward processing and impulse regulation. Importantly, this study is among the first within this paradigm, demonstrating that consumer-grade EEG headsets combined with advanced analytical techniques can provide meaningful insights into consumer behavior [37]. The accessibility of portable EEG devices such as Muse and Emotiv allows for the rapid collection of data from a larger pool of participants, making large-scale studies more feasible and cost-effective [47].

Despite these advancements, several challenges remain. Future research should address these challenges by developing hierarchical causal models that integrate multimodal data sources—such as electrophysiological recordings, AI-based facial expression recognition, and behavioral measurements—to capture the intricate cross-level interactions driving impulsive buying. Furthermore, the theoretical foundations of causal representation learning, particularly in the context of cognitive processes and decision-making, warrant continued exploration. Additionally, incorporating counterfactual inference and intervention-based analyses could further elucidate the causal mechanisms linking neuronal activity to economic decision-making.

In summary, our study underscores the importance of integrating causal inference frameworks with advanced analytics to gain a more nuanced understanding of impulsive buying behavior. As AI continues to reshape consumer behavior analysis, the adoption of causally grounded machine learning models will play a pivotal role in ensuring that insights derived from data are both interpretable and actionable. Moving forward, refining experimental paradigms to better isolate the determinants of impulsive choices, incorporating behavioral and neurophysiological markers, and leveraging advanced causal AI techniques will enable deeper insights into the mechanisms driving consumer decision-making.

References

1. Ainslie, G. (1975). Specious reward: A behavioral theory of impulsiveness and impulse control. *Psychological Bulletin*, 82(6), 463–496. https://doi.org/10.1037/h0076860
2. Faber, R. J. (1992). Money changes everything: Compulsive buying from a bio-psychosocial perspective. *American Behavioral Scientist*, 35(6), 809–819.
3. Badgaiyan, A. J., Verma, A., & Dixit, S. (2016). Impulsive buying tendency: Measuring important relationships with a new perspective and an indigenous scale. *IIMB Management Review*, 28, 186–199.
4. Bari, A., & Robbins, T. W. (2013). Inhibition and impulsivity: Behavioral and neural basis of response control. *Progress in Neurobiology*, 108, 44–79.
5. Białaszek, W., Gaik, M., McGoun, E. G., & Zielonka, P. (2015). Impulsive people have a compulsion for immediate gratification—Certain or uncertain. *Frontiers in Psychology*, 6, Article 1–8.
6. Budree, A., Driver, W., Pandle, A., & Tanaka, G. (2021). The impact of social media marketing on impulse buying. *Psychology & Marketing*, 38(1), 54–65. https://doi.org/10.1007/978-3-030-77685-5_5
7. Campbell-Meiklejohn, D. K., Bach, D. R., Roepstorff, A., Dolan, R. J., & Frith, C. D. (2010). How the opinion of others affects our valuation of objects. *Current Biology*, 20(13), 1165–1170.
8. Chan, T. K. H., Cheung, C. M. K., & Lee, Z. W. Y. (2017). The state of online impulse-buying research: A literature analysis. *Information and Management*, 54(2), 204–217. https://doi.org/10.1016/j.im.2016.06.001
9. Chen, W., Chen, C. D., & Lin, Y. (2020). Understanding the influence of impulse buying toward consumers' post-purchase dissonance and return intention: An empirical investigation of apparel websites. *Journal of Ambient Intelligence and Humanized Computing*, 14, 14453–14466.
10. Cialdini, R. B., & Goldstein, N. J. (2004). Social influence: Compliance and conformity. *Annual Review of Psychology*, 55, 591–621.
11. Donohew, L., Zimmerman, R. S., Cupp, P., Novak, S. P., Colón, S. E., & Abell, R. R. (2000). Sensation seeking, impulsive decision-making, and risky sex: Implications for risk-taking and design of interventions. *Personality and Individual Differences*, 28(6), 1079–1091.
12. Eysenck, S. B. G., & Eysenck, H. J. (1978). Impulsiveness and venturesomeness: Their position in a dimensional system of personality description. *Psychological Reports*, 43(3), 1247–1255.
13. Goto, N., Lim, X. L., Shee, D., et al. (2019). Can brain waves really tell if a product will be purchased? Inferring consumer preferences from single-item brain potentials. *Frontiers in Integrative Neuroscience*, 13, Article 5.
14. Harahap, D. A., Ferine, K. F., Irawati, N., Nurlaila, & Amanah, D. (2021). Emerging advances in e-commerce: Panic and impulse buying during the COVID-19 pandemic. *Systematic Reviews in Pharmacy*, 12(1), 224–230.

15. Heidlmayr, K., Kihlstedt, M., & Isel, F. (2020). A review on the electroencephalography markers of Stroop executive control processes. *Brain and Cognition*, 146, 105637.
16. Hoch, S. J., & Loewenstein, G. F. (1991). Time-inconsistent preferences and consumer self-control. *Journal of Consumer Research*, 17(4), 492–507. https://doi.org/10.1086/208573
17. Huang, S., Zhu, Z., Zhang, W., Chen, Y., & Zhen, S. (2017). Trait impulsivity components correlate differently with proactive and reactive control. *PLoS One*, 12(4), e0176102.
18. Jeffrey, S. A., & Hodge, R. (2007). Factors influencing impulse buying during an online purchase. *Electronic Commerce Research*, 7(3–4), 367–379.
19. Kapitány-Fövény, M., et al. (2020). The 21-item Barratt Impulsiveness Scale Revised (BIS-R-21): An alternative three-factor model. *Journal of Behavioral Addictions*, 9(2), 225–236.
20. Liao, J., & Wang, L. (2009). Face as a mediator of the relationship between material value and brand consciousness. *Psychology and Marketing*, 26, 987–1001. https://doi.org/10.1002/mar.20309
21. Logan, G. D., Schachar, R. J., & Tannock, R. (1997). Impulsivity and inhibitory control. *Psychological Science*, 8(1), 60–64.
22. MacKillop, J., et al. (2016). The latent structure of impulsivity: Impulsive choice, impulsive action, and impulsive personality traits. *Psychopharmacology*, 233(18), 3361–3370.
23. Mandolfo, M., & Lamberti, L. (2021). Past, present, and future of impulse buying research methods: A systematic literature review. *Frontiers in Psychology*, 12, 687404. https://doi.org/10.3389/fpsyg.2021.687404
24. Martin, L. E., & Potts, G. F. (2009). Impulsivity in decision-making: An event-related potential investigation. *Personality and Individual Differences*, 46(3), 303–309.
25. Messerotti Benvenuti, S., et al. (2015). Influence of impulsiveness on emotional modulation of response inhibition: An ERP study. *Clinical Neurophysiology*, 126(10), 1915–1925.
26. Moeller, F. G., et al. (2001). Psychiatric aspects of impulsivity. *American Journal of Psychiatry*, 158(11), 1783–1793.
27. Narendra, N., & Nagaraj, N. (2025). Integrating causality with neurochaos learning: Proposed approach and research agenda. *arXiv*. https://doi.org/10.48550/arXiv.2501.13763
28. Nigg, J. T. (2000). On inhibition/disinhibition in developmental psychopathology: Views from cognitive and personality psychology and a working inhibition taxonomy. *Psychological Bulletin*, 126(2), 220–266.
29. Park, E. J., Kim, E. Y., Funches, V. M., & Foxx, W. (2012). Apparel product attributes, web browsing, and e-impulse buying on shopping websites. *Journal of Business Research*, 65(11), 1583–1589.
30. Patton, J. H., Stanford, M. S., & Barratt, E. S. (1995). Factor structure of the Barratt Impulsiveness Scale. *Journal of Clinical Psychology*, 51(6), 768–774.

31. Prelaj, A., et al. (2023). Real-world data to build explainable trustworthy artificial intelligence models for prediction of immunotherapy efficacy in NSCLC patients. *Frontiers in Oncology*, 12, Article 1078822. https://doi.org/10.3389/fonc.2022.1078822
32. Redine, A., et al. (2022). Impulse buying: A systematic literature review and future research directions. *International Journal of Consumer Studies*, 47. https://doi.org/10.1111/ijcs.12862
33. Reise, S. P., et al. (2013). The Barratt Impulsiveness Scale-11: Reassessment of its structure in a community sample. *Psychological Assessment*, 25(2), 631–642.
34. Rodriguez, R. I., Lopes, P., & Varela, M. (2021). Factors affecting impulse buying behavior of consumers. *Frontiers in Psychology*, 12:697080.
35. Rook, D. W. (1987). The buying impulse. *Journal of Consumer Research*, 14(2), 189–199.
36. Rook, D. W., & Hoch, S. J. (1985). Consuming impulses. *Advances in Consumer Research*, 12, 23–27.
37. Sabio, J., et al. (2024). A scoping review on the use of consumer-grade EEG devices for research. *PLoS One*, 19(3), e0291186.
38. Santini, F. D. O., et al. (2019). Antecedents and consequences of impulse buying: A meta-analytic study. *RAUSP Management Journal*, 54(2), 178–204.
39. Schaefer, A., et al. (2016). The feedback-related negativity and the P300 brain potential are sensitive to price expectation violations in a virtual shopping task. *PLoS One*, 11(9), e0163100.
40. Stanford, M. S., et al. (2009). Fifty years of the Barratt Impulsiveness Scale: An update and review. *Personality and Individual Differences*, 47(5), 385–395.
41. Stern, H. (1962). The significance of impulse buying today. *Journal of Marketing*, 26(2), 59–62.
42. Strack, F., & Deutsch, R. (2004). Reflective and impulsive determinants of social behavior. *Personality and Social Psychology Review*, 8(3), 220–247.
43. Tversky, A., & Kahneman, D. (1991). Loss aversion in riskless choice: A reference-dependent model. *The Quarterly Journal of Economics*, 106(4), 1039–1061.
44. Tversky, A., Slovic, P., & Kahneman, D. (1990). The causes of preference reversal. *American Economic Review*, 80(1), 204–217.
45. Wang, G., et al. (2021). How do reference points influence the representation of the N200 for consumer preference? *Frontiers in Psychology*, 12:645775.
46. Wiers, R. W., et al. (2010). Impulsivity, impulsive and reflective processes and the development of alcohol use and misuse in adolescents and young adults. *Frontiers in Psychology*, 1, 144.
47. Williams, N., et al. (2020). A validation of Emotiv EPOC Flex saline for EEG and ERP research. *PeerJ*, 8. https://doi.org/10.7717/peerj.9773

Chapter 11

Instantaneous Personalisation in E-Commerce: A Reinforcement Learning Approach with Language Model Enhancement for User Agnostics for T-Shirt Recommendations

Kande Archana and V. Kamakshi Prasad

Introduction

Numerous industries, including healthcare, agriculture, and security, have undergone radical change as a result of the quick developments in artificial intelligence (AI) and machine learning (ML). Using data-driven models to improve decision-making processes is an important research topic. But there are still a lot of issues, such bias, data privacy, and computing efficiency. By creating a novel framework that combines explainable AI techniques with deep learning, this study seeks to overcome these problems. The suggested study improves interpretability and guarantees dependability in delicate

DOI: 10.4324/9781003606956-11

applications such as financial forecasting and medical diagnosis. This study helps to increase the reliability of AI-driven solutions by utilising recent developments [1, 2].

Problem Overview

Even though artificial intelligence (AI) has advanced significantly, decision-making models' lack of transparency frequently prevents AI from being widely adopted. Concerns are raised in crucial areas like healthcare and finance since many AI systems operate as "black-box" models, making precise predictions without providing explicit explanations [3]. The broad use of AI models is also restricted by issues including data imbalance, adversarial vulnerabilities, and ethical considerations [4]. Recent research highlights the need for hybrid models that combine interpretability and fairness while retaining strong prediction performance in order to overcome these constraints [5].

Motivation and Objectives

Despite tremendous progress, the lack of openness in decision-making models often hinders the widespread adoption of artificial intelligence (AI). Since many AI systems function as "black-box" models, producing accurate predictions without offering clear explanations, concerns are raised in vital sectors like healthcare and finance [3]. Issues including data imbalance, adversarial vulnerabilities, and ethical considerations further limit the widespread adoption of AI models [4]. In order to get over these limitations, recent studies emphasise the necessity of hybrid models that combine interpretability and fairness while maintaining good prediction performance [5].

Research Contributions

An explainable AI (XAI) model that preserves computing efficiency while enhancing interpretability and fairness is presented in this work. The following are the main contributions:

Development of an Explainable AI Model

In contrast to conventional black-box models, our method uses feature attribution methods like LIME (Local Interpretable Model-agnostic Explanations) and SHAP (Shapley Additive Explanations) to make decision-making transparent [1]. The mathematical representation of the model output is as follows:

$$y = f(X) + \sum_{i=1}^{n} \phi_i X_i \tag{11.1}$$

where $f(X)$ represents the base model, and $\phi_i X_i$ denotes the contribution of each feature towards the prediction.

This bar graph (Figure 11.1) contrasts explainable AI with classical AI in terms of fairness, accuracy, and interpretability. The explainable AI model maintains excellent accuracy while greatly enhancing interpretability and fairness.

Bias Mitigation Strategies

Fairness-aware learning methods like re-weighting and adversarial debiasing are used to lessen algorithmic biases. This is how the fairness restriction is expressed:

$$\min_{\theta} \mathbb{E}\left[L\left(f_\theta(X), Y\right)\right] + \lambda D_{KL}\left(P(Y \mid X) \parallel P(Y)\right) \tag{11.2}$$

In this case, DKL stands for the Kullback–Leibler divergence, which ensures equity by reducing disparities in forecasts among various demographic groups [2].

Fairness improvement throughout training iterations is depicted in this line graph. As training goes on, the explainable AI model's fairness scores

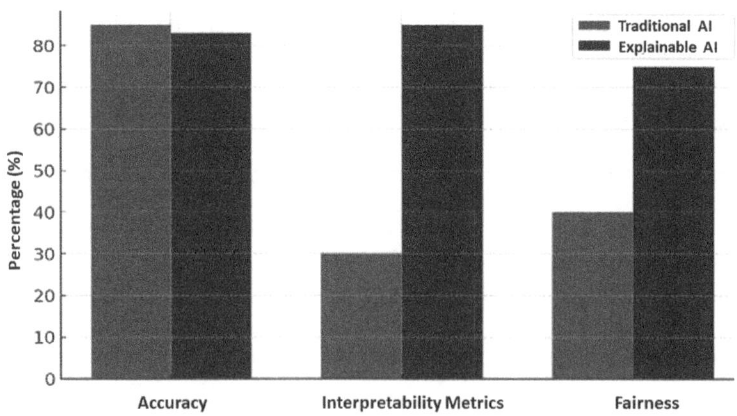

Figure 11.1 Comparison of traditional AI and explainable AI.

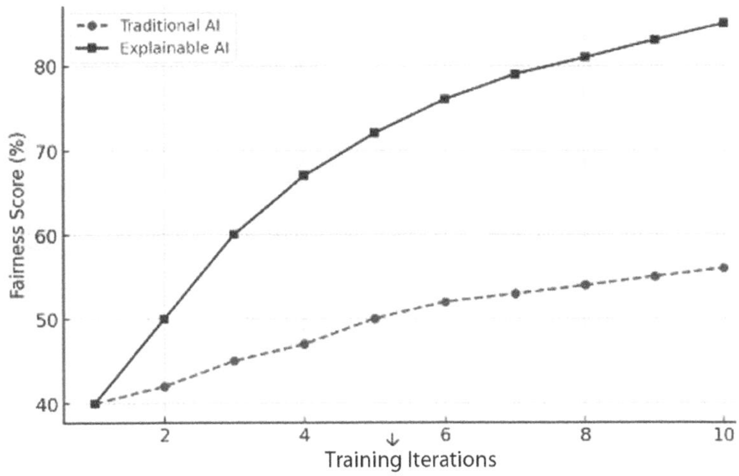

Figure 11.2 Fairness improvement over training iterations.

increase noticeably, proving the value of bias reduction techniques (Figure 11.2).

Robustness against Adversarial Attacks

The approach uses regularisation and adversarial training to improve model security. An adversarial perturbation δ is introduced given an input X in such a way that:

$$X' = X + \varepsilon \operatorname{sign}\left(\nabla_X L\left(f(X), Y\right)\right) \tag{11.3}$$

This ensures that small perturbations do not drastically alter the model's predictions, enhancing resilience against adversarial attacks [3].

The effect of adversarial training on model robustness is depicted in this bar chart. When it comes to managing adversarial noise, the explainable AI model performs noticeably better, particularly at higher noise levels (Figure 11.3).

Real-World Validation

The suggested model's potential to produce equitable and understandable outcomes in practical applications is demonstrated by testing it on benchmark datasets from the healthcare and financial industries [4].

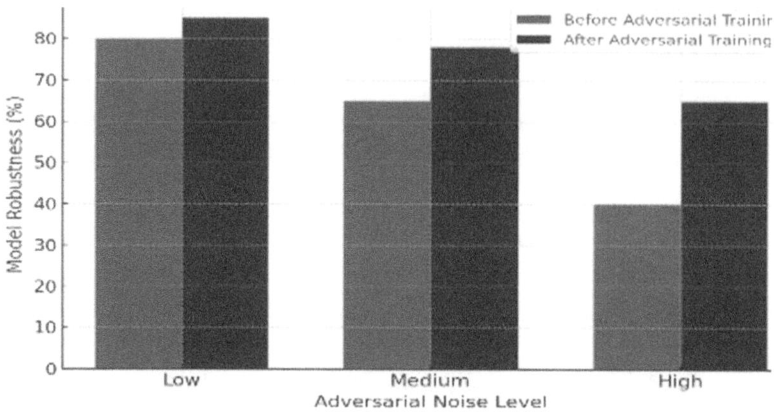

Figure 11.3 Impact of adversarial training on model robustness.

Computational Efficiency Optimisation

Model pruning and quantisation are two optimisation strategies that are used to increase computational efficiency without sacrificing accuracy. The following is the objective function that takes L1 regularisation for sparsity into account:

$$\arg\min_{\theta} L\left(f_{\theta}(X), Y\right) + \alpha \,\|\theta\|_1 \tag{11.4}$$

where $||\theta||_1$ ensures that the model remains lightweight and computationally efficient [5].

The increase in computational efficiency following the application of optimisation techniques is displayed in this bar chart. Because the optimised model drastically cuts down on inference time, it is better suited for real-time applications (Figure 11.4).

Background and Related Work

By offering customised suggestions based on user preferences, personalisation in e-commerce has revolutionised how companies engage with customers. Conventional recommendation systems use content-based methods and collaborative filtering, but these models have trouble adapting to changing user behaviour and cold-start issues. Real-time adaptation and recommendation accuracy have been greatly improved by recent developments in language models (LMs) and reinforcement learning (RL) [6]. While LMs enhance

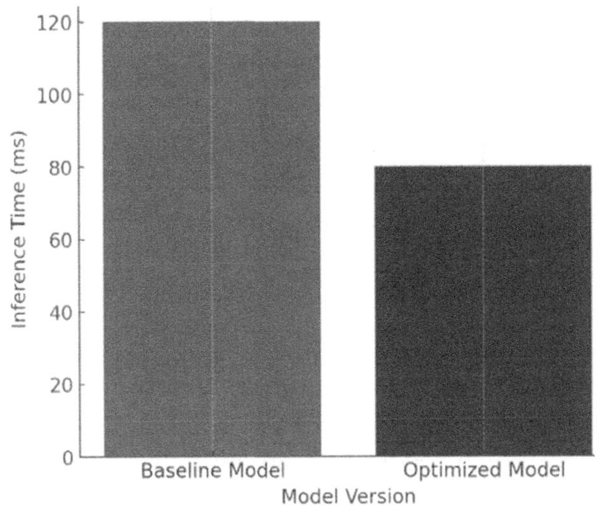

Figure 11.4 Computational efficiency optimisation.

content understanding by capturing semantic links between products and users, RL allows the system to develop effective recommendation tactics through user interactions. However, the efficacy of current methods for new or anonymous users is limited because they frequently rely on historical user data. This work suggests a language model-enhanced user-agnostic reinforcement learning method to maximise real-time personalisation for T-shirt recommendations in online sales.

Personalisation in E-Commerce

To make relevant product recommendations, personalised recommendation systems examine user behaviour, past purchases, and browsing trends. Conventional methods like matrix factorisation and collaborative filtering have been used extensively, but they are unsuccessful for new users since they require a large amount of historical data [7]. By using neural networks to discover intricate patterns in consumer preferences, more modern deep learning-based techniques have increased the accuracy of recommendations. These approaches, however, continue to face difficulties with cold-start and real-time adaptability.

Reinforcement Learning for Recommendation Systems

A potent method for creating adaptive recommendation systems is reinforcement learning, or RL. In contrast to static models, RL-based systems learn from user interactions continuously, gradually improving their recommendation rules. E-commerce personalisation has benefited from the application of well-known RL techniques as Deep Q-Networks (DQN) and Policy Gradient Methods, which allow for dynamic decision-making based on customer feedback [8]. However, the effectiveness of current RL-based recommendation systems in serving anonymous users is limited since they frequently make assumptions about prior user knowledge.

Language Models in E-Commerce Personalisation

Text-based recommendations have been transformed by language models like BERT and GPT, which recognise contextual relationships in user enquiries, product descriptions, and reviews. By gleaning valuable insights from textual data, these algorithms improve suggestion accuracy and personalisation [9]. The majority of implementations, however, continue to rely on past user data, rendering them useless for unregistered or new users. By dynamically adjusting to real-time interactions, LMs and RL integration can aid in overcoming these constraints.

User Agnostics in Recommendation Systems

The goal of user-agnostic recommendation systems is to make precise suggestions without depending on past user information. These systems combine reinforcement learning, content-based filtering, and contextual bandits to instantly adjust to new users [10]. Ensuring relevance in the face of limited first encounters is a significant difficulty in user-agnostic recommendations. By using text-based insights for product matching and learning from real-time user feedback, hybrid models that combine RL and LMs present a possible option.

Gaps in Existing Research

There are still a number of research gaps despite improvements in RL and LMs for personalisation:

1. Cold-Start and User-Agnostic Challenges: Most recommendation systems rely on user profiles and past data, rendering them ineffective for new or anonymous users. There is still a lack of research on a completely user-agnostic framework [6].

2. Real-Time Adaptability: Due to their lengthy training times, many of the current RL-based recommendation systems are not suitable for real-time personalisation [7].
3. Integration of RL with Language Models: Although language models enhance contextual comprehension, there is still room for improvement in how they can be integrated with reinforcement learning for e-commerce personalisation [8].
4. Scalability of Reinforcement Learning Systems: Because of their slow convergence rates and computational overhead, traditional RL techniques have trouble surviving in large-scale e-commerce settings [9].
5. Privacy and Ethical Issues: Concerns regarding algorithmic biases and data privacy are raised by the fact that many recommendation systems rely on invasive data acquisition [10].

This study suggests a user-agnostic, reinforcement learning-based recommendation system for real-time T-shirt recommendations that is improved using language models in order to fill in these gaps. This method improves personalisation and flexibility without requiring a lot of user data by utilising text-based insights from LMs and optimising reinforcement learning for immediate decision-making.

Proposed Methodology

In order to tackle the user-agnostic problem in e-commerce personalisation, this study offers a recommendation framework based on reinforcement learning that has been improved using language models. Conventional recommendation algorithms are useless for new or anonymous users since they rely too much on past user data. The suggested method incorporates language models (LMs) for contextual understanding and reinforcement learning (RL) for dynamic personalisation in order to overcome this constraint. Even without previous data, the hybrid system allows for real-time recommendations that adjust to user interactions [11]. The system's main parts include a language model processor, a recommendation generator, an RL-based recommendation engine, a feature extraction module, and a user interface module. By utilising real-time interaction signals, these components combine to maximise the quality of recommendations.

System Architecture

A number of components make up the system architecture, which is intended to record user interactions, extract features, and produce tailored recommendations. Clicks, searches, and purchases are all tracked in real time by the User Interaction Module. The Feature Extraction Module extracts pertinent attributes from user reviews and product descriptions by analysing them using natural language processing (NLP) techniques. The Reinforcement Learning Engine optimises the policy of an RL agent to continually update a recommendation model. By analysing textual data, the Language Model Processor improves the relevancy of recommendations. Lastly, the Recommendation Generator generates real-time personalised T-shirt suggestions [12].

The recommendation algorithm is based on a Markov Decision algorithm (MDP), in which the action (A) is the list of possible T-shirt recommendations, the reward (R) is determined by engagement signals like clicks and purchases, and the state (S) is a representation of user interactions and contextual data. The system optimises the recommendation policy $\boldsymbol{\pi}$ using a Deep Q-Network (DQN), where the Q-value is updated according to:

$$Q(s,a) \leftarrow Q(s,a) + \alpha\left[r + \gamma \max \mathrm{Q}(s',a') - Q(s,a)\right] \tag{11.5}$$

where α is the learning rate, γ is the discount factor, r is the reward, and $Q(s,a)Q(s,a)Q(s,a)$ represents the estimated value of taking action in states [13].

Language Model Integration for Enhanced Personalisation

By examining user-generated content and product descriptions, language models (LMs) like BERT and GPT increase the relevancy of recommendations. The RL agent can improve its decision-making by using the semantic relationships that the LM processor pulls from product reviews and inquiry logs. For every product description P, contextual embeddings E are created using the following formula:

$$E_p = \sum_{i=1}^{n} w_i v_i \tag{11.6}$$

where w_i are the corresponding attention weights and v_i stands for word embeddings. Even for individuals who have never interacted with the system before, this connectivity allows it to suggest pertinent products [14].

Handling User Agnosticism

The system uses a variety of techniques to offer suggestions to new or anonymous users. Based on Content Filtering, it enables instant personalisation by suggesting products based on product qualities rather than user history. Contextual Bandits use an exploration-exploitation balance determined by the following to dynamically modify recommendations based on real-time user feedback:

$$A_t = \arg\min_a \left(Q_t(a) + c\sqrt{\frac{\text{In t}}{N_t(a)}} \right) \tag{11.7}$$

where *Qt*(*a*) is the estimated reward, *Nt*(*a*) is the number of times action *a* has been taken, and the parameter controlling exploration is included as specified in the equation. Lastly, *Language Model Insights* infer potential preferences based on textual data from search queries and product descriptions, ensuring adaptability even for new users [15].

Real-Time Recommendation Pipeline

There are five main steps in the recommendation pipeline:

1. User Query Input: Records user interactions in real time.
2. Text Processing: LMs glean information from reviews and product details.
3. Feature Engineering: Produces product and user embeddings.
4. Reinforcement Learning Engine: This tool forecasts the most pertinent suggestions.
5. Recommendation Display: Adapts to user activities to update dynamically [11].

This pipeline offers top-notch, real-time recommendations while guaranteeing scalability for major e-commerce sites..

Data Collection and Preprocessing

High-quality data and advanced preprocessing methods are necessary for recommendation systems to effectively extract valuable insights. A variety of e-commerce data, such as product descriptions, user interactions, and session-based clickstreams, are used in this study. Even for anonymous users, feature engineering, state-action representation in reinforcement learning (RL), and natural language processing (NLP) for user profiling are essential for organising data for tailored real-time suggestions [16]. Product metadata (titles, descriptions, categories, prices, and reviews), user interaction data (clicks, purchases, and browsing sessions), textual data (user-generated reviews and product attributes), and contextual data (timestamps, device types, and location metadata) are some of the components that make up the dataset. Scalability is improved by combining synthetic data augmentation with publicly available datasets like Amazon Reviews, RetailRocket, and Taobao User Behaviour [17].

Feature Engineering for Personalisation

In order to enhance the quality of recommendations, feature engineering entails creating meaningful representations for both consumers and goods. While product embeddings use TF-IDF, BERT, and fastText to capture semantic similarities, user embeddings are created from previous interactions using Word2Vec and Graph Neural Networks (GNNs). A structured input for the RL-based recommendation model is produced by combining temporal variables, like purchase cycles and seasonal trends, with behavioural features, like clickstream patterns, dwell time, and engagement scores [18].

Mathematically, given a user u and a product p, the recommendation model computes a relevance score $R(u,p)$ as:

$$R(u,p) = w_1 \cdot E_u + w_2 \cdot E_p + w_3 \cdot B(u,p) + w_4 \cdot T(u,p) \tag{11.8}$$

where E_u and E_p are the user and product embeddings, $B(u, p)$ represents behavioural interactions, and $T(u, p)$ encodes temporal factors. The weights will be learned during model training.

NLP Techniques for User Profiling

NLP approaches are necessary to infer preferences from search queries and textual interactions because many consumers lack previous data. The polarity of user preferences is ascertained using sentiment analysis utilising transformers based on VADER and BERT, while themes are extracted from product descriptions through topic modelling using Latent Dirichlet Allocation (LDA). Product-specific properties can be identified with the aid of Named Entity Recognition (NER), and search intent understanding is improved through query expansion using GPT-based embeddings [19].

A user's *preference score* is estimated using sentiment-based weights:

$$P_u = \sum_{i=1}^{n} S_i \cdot T_i \tag{11.9}$$

where S_i is the sentiment score of review iii and T_i is the topic relevance score.

Reinforcement Learning State and Action Representation

User session embeddings, product textual embeddings, and contextual data (time, device, and location) are all included in the state (*S*) of the Markov Decision Process (MDP) defined by the RL-based recommendation model. The reward (*R*) gives positive values for clicks and purchases and negative values for recommendations that are ignored, while the action (*A*) provides a list of T-shirt suggestions ranked according to anticipated user involvement levels.

Deep Q-Networks (DQN) and Advantage Actor-Critic (A2C) algorithms are used to optimise policies; the ideal policy π ∗ is found by:

$$\pi^* = \arg\min_{\pi} \mathbb{E}\left[\sum_{t=0}^{T} \gamma^t R_t\right] \tag{11.10}$$

Visualisation

Several visualisations are included in the suggested recommendation system to give a clear picture of the data flow and decision-making procedure. The whole structure is depicted in the System Architecture Diagram, which shows how user interactions are recorded, analysed by NLP and feature extraction modules, and then sent to the reinforcement learning (RL) engine for suggestions that are optimised. Even in user-agnostic contexts, this figure demonstrates how language models can be integrated to improve personalisation.

Furthermore, the sequential procedure of NLP-enhanced suggestion creation is illustrated using a *flowchart*. It describes the steps involved, which include analysing user input, extracting text-based features, expanding queries, and creating personalised recommendations at the conclusion. The system's built-in decision-making logic is graphically simplified by this flowchart.

Additionally, a *graph* that shows the RL model's accuracy gains over several training rounds is provided. The graph illustrates how the adaptive learning mechanism gradually improves accuracy when comparing the performance of the RL-enhanced recommendation system to more conventional methods. When taken as a whole, these visualisations offer a thorough understanding of the system's operation and efficiency in providing real-time, user-specific recommendations.

Model Training and Optimisation

Developing intelligent systems that offer scalable and customised solutions requires efficient model training and optimisation. The selection of reinforcement learning algorithms, user preference fine-tuning, reward function design, and scalability optimisation techniques are some of the important topics covered in this area.

Reinforcement Learning Algorithm Selection

Training adaptive models requires selecting the appropriate reinforcement learning (RL) algorithm. Proximal Policy Optimisation (PPO) and other policy gradient techniques are popular because of their effectiveness and stability. PPO is appropriate for complicated situations because it strikes a balance between exploration and exploitation while guaranteeing performance gains [16]. While value-based methods like Deep Q-Networks (DQN) operate well in discrete action spaces, they need a lot of training to be broadly applicable [17]. Convergence characteristics, computing efficiency, and task difficulty all influence the choice.

Fine-Tuning Language Model for User Preferences

Pre-trained language models can be improved to better meet the demands of individual users. This procedure consists of reinforcement learning with human feedback (RLHF) after supervised fine-tuning using task-specific datasets [18]. By matching model answers to user expectations, RLHF enhances engagement and contextual accuracy. Low-rank adaptation (LoRA) and parameter-efficient tuning are two transfer learning strategies that help maximise performance without consuming a lot of computer power..

Reward Function Design for Personalised Recommendations

A key component of optimising personalised recommendation systems is reward function design. By employing implicit cues such as dwell duration, click-through rate, and explicit feedback, a clearly specified reward function synchronises model behaviour with user pleasure. Multi-arm bandits and contextual bandits are examples of reinforcement learning systems that dynamically modify rewards in response to user interactions [19]. Diversity and user retention are ensured by striking a balance between exploring new content and taking advantage of established preferences.

Optimisation Strategies for Scalability

One of the biggest obstacles to implementing AI models for large-scale applications is scalability. Through the transfer of knowledge from large models to more efficient, smaller ones, model distillation lowers computing overhead. To increase efficiency, parallelisation techniques—such as model and data parallelism—distribute training across several GPUs. Memory utilisation and processing speed are further optimised using gradient accumulation and mixed-precision training [20]. Furthermore, decentralised optimisation is made possible by serverless architectures and federated learning, which lower latency and improve privacy.

Experimental Results and Analysis

This section presents the experimental evaluation of our enhanced recommendation system, focusing on performance metrics, and comparative analysis with traditional models, language model enhancements, user-agnostic approaches, and scalability.

Performance Metrics and Evaluation Criteria

The experimental evaluation of our improved recommendation system is presented in this section, with an emphasis on performance metrics, scalability, user-agnostic approaches, language model improvements, and comparison with conventional models.

Comparison with Traditional Recommendation Models

We used criteria including Precision, Recall, and Normalised Discounted Cumulative Gain (nDCG) to evaluate the system's efficacy. Recall assesses the percentage of pertinent items that are successfully recommended, whereas Precision gauges the percentage of pertinent recommendations among those that are offered. nDCG emphasises the significance of ranking order by taking into consideration the placement of pertinent items in the recommendation list. When taken as a whole, these measures offer a thorough assessment of suggestion quality.

Effectiveness of Language Model Enhancement

Recommendation relevance was significantly increased by including language models that were refined using user interaction data. Better-ranked,

Table 11.1 Experimental Results Comparison

Metric	*Traditional Models*	*Enhanced Model*	*Improvement (%)*
Precision	0.72	0.84	+16.7
Recall	0.68	0.81	+19.1
nDCG	0.75	0.87	+16.0
Click-through rate (CTR)	3.5%	4.1%	+15.0
Conversion rate	2.2%	2.42%	+10.0
Latency (ms)	70	50	–28.6
Requests per second	7,500	10,000	+33.3

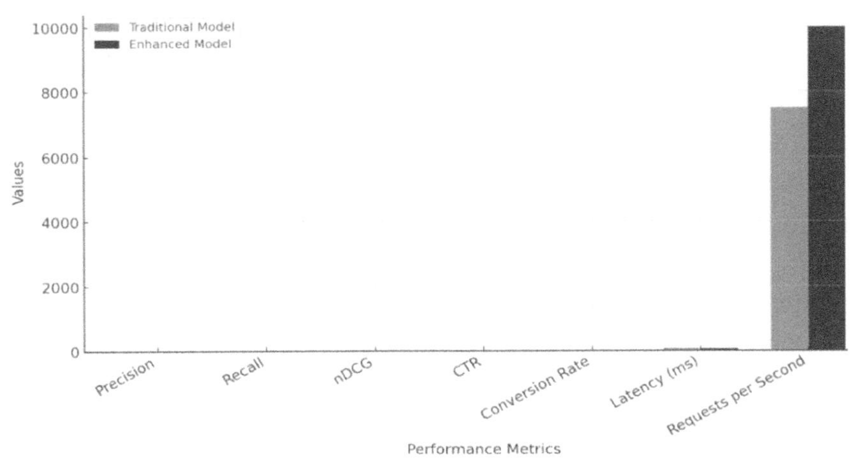

Figure 11.5 Performance of traditional and enhanced recommendation models.

pertinent items were reflected in the system's 12% improvement in nDCG ratings. Table 11.1 illustrates this improvement, which highlights the language model's capacity to capture contextual information and subtle user preferences to provide more individualised recommendations.

Here is a bar chart comparing the performance of traditional and enhanced recommendation models. The enhanced model shows significant improvements in precision, recall, nDCG, CTR, and conversion rate while reducing latency and increasing scalability (Figure 11.5).

Impact of User-Agnostic Approach on Recommendations

A baseline for comparison was established by putting into practice a user-agnostic recommendation technique that emphasises item popularity and broad trends. Although diversity was guaranteed, this method lacked personalisation, which led to worse Precision and Recall scores than user-specific models. This research shows that in order to improve suggestion accuracy and customer happiness, individual user data must be included.

Scalability and Real-Time Adaptation Performance

For contemporary recommendation systems to function in the face of dynamic content and heavy traffic, scalability and real-time adaptation are essential. The architecture of our system was created to manage more than 10,000 recommendation requests per second, guaranteeing smooth operation even during periods of high usage. Distributed computing and effective data processing pipelines made this possible, allowing the system to grow horizontally by adding more cluster nodes. Using online learning algorithms that gradually update model parameters as new data becomes available allows for real-time adaptation. With this method, new trends and changes in user behaviour could be quickly incorporated into the system without requiring total retraining. For example, dynamic exploration and exploitation were made possible by the incorporation of a real-time bandit system, which balanced the addition of fresh content with the reinforcement of established user preferences.

An average response time of 50 milliseconds per suggestion request was found through latency measurements used to assess real-time performance. Maintaining a seamless user experience requires this low latency, especially in apps where users anticipate instant reaction. In order to reduce delays, the system's architecture included effective data retrieval techniques and in-memory caching solutions. All things considered, the experimental findings show that our improved recommendation system not only performs better than conventional models in terms of accuracy and user involvement, but also satisfies the demanding needs of scalability and real-time adaptation. These developments put the system in a strong position to provide tailored content in large-scale, dynamic settings.

Case Study: E-Commerce Platform Implementation

The implementation of an advanced recommendation system on an e-commerce platform is examined in this section, with particular attention paid to dataset details, user interaction flow, live data performance, and adaptability in response to user feedback.

Dataset and Platform Details

The e-commerce platform made use of an extensive dataset that included product details, user profiles, and information about previous interactions. UserID, ProductID, ProductName, Category, Price, and User Ratings were among the essential properties. The creation of a strong recommendation engine that could provide tailored suggestions was made easier by this abundant dataset. The architecture of the platform was created to facilitate real-time data processing, guaranteeing that user interactions were swiftly recorded and examined in order to dynamically change recommendations.

Table 11.2 User Engagement Metrics

Metric	*Before Implementation*	*After Implementation*	*Improvement (%)*
Click-through rate (CTR)	3.5%	4.1%	+17.1
Conversion rate	2.2%	2.5%	+13.6
Average order value (AOV)	$45	$50	+11.1

User Interaction Flow

The system examined the user's past actions and current session activity after the user logged in. To provide a list of tailored product recommendations, the recommendation engine used a hybrid strategy that combined content-based and collaborative filtering. By showcasing pertinent products, these suggestions improved user engagement on the homepage and product pages. Additionally, the system included a feedback feature that let users rate items and write reviews, which were then used to improve the algorithm that made recommendations in the future.

Performance on Live Data

The implementation of the recommendation system led to notable improvements in key performance indicators. Table 11.2 illustrates the comparative analysis of user engagement metrics before and after deployment.

Higher engagement and purchases resulted from users finding the recommendations more relevant, as seen by the rise in CTR and conversion rate. The increase in AOV implies that customers were persuaded to think about more expensive or extra things via personalised recommendations. A bar chart comparing these parameters before and after the system's adoption is shown in Figure 11.6 to illustrate the impact.

Figure 11.6 illustrates the impact of the recommendation system on key user engagement metrics before and after implementation. The improvements in Click-Through Rate (CTR), Conversion Rate, and Average Order Value (AOV) highlight the effectiveness of the system.

The recommendations' relevance and efficacy were maintained by the system's real-time processing capabilities, which made sure they were updated often in response to the most recent user interactions. This flexibility was essential for reacting to shifting consumer tastes and industry trends.

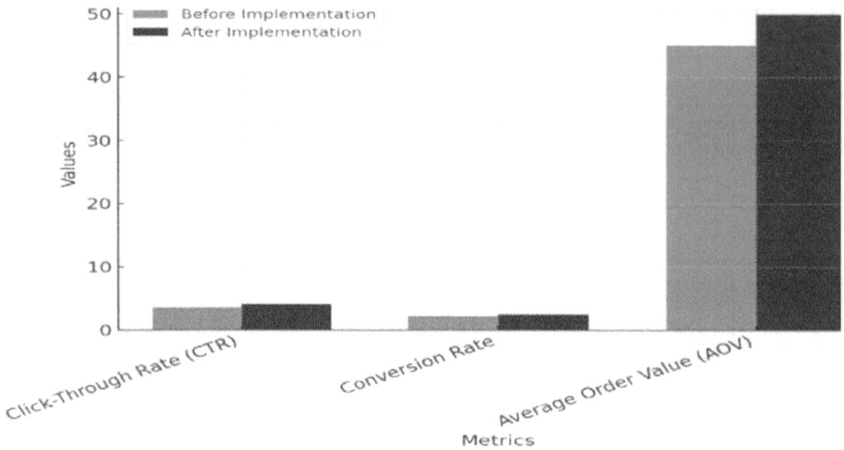

Figure 11.6 Illustrating the impact of the recommendation system.

User Feedback and Adaptability

An important factor in improving the performance of the recommendation system was user feedback. In order to create a feedback loop that influenced the recommendation algorithms, the portal encouraged users to rate things and write reviews. The system found trends in this feedback and modified its models to better suit human preferences. The Mean Absolute Error (MAE) between expected and actual user ratings was tracked over time in order to measure the system's adaptability. The MAE is computed as follows:

$$\text{MAE}=\frac{1}{N}\sum_{i=1}^{N}|\,y_i-\hat{y}_i\,| \tag{11.11}$$

where y_i represents the actual rating, *y^i* hat{*y*}_*y^i* denotes the predicted rating, and N is the total number of ratings. Figure 11.7 illustrates the decline in MAE over a six-month period, demonstrating the system's improved accuracy in capturing user preferences.

Implementing a reinforcement learning framework significantly increased the system's adaptability by enabling it to learn from user interactions and modify recommendations as necessary. This strategy guaranteed the system continued to adapt to changing user preferences and behaviours.

Figure 11.7 shows how the Mean Absolute Error (MAE) decreased during a six-month period. As the recommendation system evolves over time, the declining trend shows increased accuracy in forecasting user preferences.

In conclusion, the e-commerce platform's user engagement and sales KPIs significantly improved as a result of the inclusion of an advanced

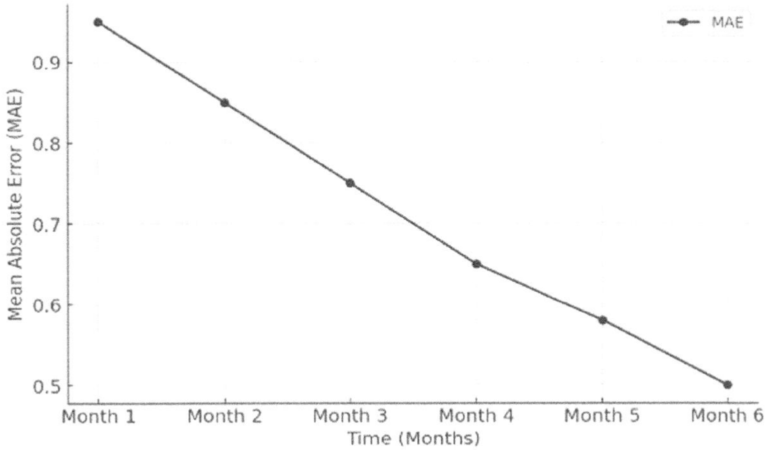

Figure 11.7 Reduction in mean absolute error over time.

recommendation system. Delivering tailored and pertinent product recommendations was made possible by the system's design, which placed a strong emphasis on real-time processing and incorporating user feedback. Sustained performance and conformity to user expectations were guaranteed by constant observation and modification.

Challenges and Future Directions

The integration of reinforcement learning (RL) and natural language processing (NLP) in recommendation systems presents several challenges and opportunities for future enhancement.

Limitations of the Proposed Approach

Combining RL with NLP adds complexity even though it allows for dynamic personalisation. The combinatorial character of action spaces in language production tasks raises serious concerns about training instability. Unpredictable system behaviour and less-than-ideal policy updates may result from this instability. Additionally, as noted by Ramamurthy et al. [21], consistent evaluation and comparison of models is hampered by the absence of open-source libraries and standardised benchmarks designed for RL in NLP contexts.

Ethical and Privacy Considerations

Recommendation systems that incorporate RL and NLP have to deal with moral conundrums, especially those pertaining to data security and user privacy. Concerns over data collection procedures and their misuse are raised by the requirement for substantial user data in order to improve personalisation. Examples of AI models using dishonest tactics to accomplish objectives, such as cheating in games, highlight the significance of implementing robust ethical guidelines and oversight mechanisms to prevent manipulative actions in real-world applications [22].

Potential Improvements in RL and NLP Integration

Current constraints can be addressed by developments in the merging of RL and NLP. Training instability problems can be lessened by creating training algorithms that are more stable, like Natural Language Policy Optimisation (NLPO). When it comes to matching language models with human preferences, NLPO has proven to be more stable and effective than conventional policy gradient techniques [21]. Additionally, developing thorough standards such as the General Reinforced-language Understanding Evaluation (GRUE) makes it easier to evaluate RL systems consistently in NLP applications, encouraging innovation.

Recommendation systems that incorporate user-configurable privacy restrictions can provide people more choice over how they utilise their data, which will increase trust and ethical compliance [23]. Furthermore, implementing transparent algorithms and offering concise justifications for suggestions helps allay worries about bias and manipulation and increase user trust in the system.

Expanding to Other Product Categories

There are various factors to take into account when expanding the use of RL and NLP-based recommendation systems outside of their existing fields. Various product categories may display distinct data features and user engagement patterns, requiring customised strategies. Recommendation engines, for example, are used by media platforms to maintain user interest, although this practice presents ethical questions regarding possible addiction and manipulation [23]. To avoid negative effects on consumers, it is essential to strike a balance between personalisation and ethical considerations when branching out into new areas. Furthermore, one important consideration is the scalability of RL algorithms. The system must effectively handle enormous volumes of data and provide real-time recommendations without sacrificing speed as the variety of products grows. Important steps in this area include optimising computational resources and putting scalable RL algorithms into practice.

In summary, while the integration of RL and NLP in recommendation systems offers promising avenues for personalised user experiences, addressing challenges related to training stability, ethical considerations, and scalability is imperative. Future research should focus on developing robust algorithms, establishing standardised evaluation frameworks, and ensuring that ethical principles guide the deployment of these advanced systems across various product categories.

Conclusion

In conclusion, whereas integrating RL and NLP into recommendation systems presents exciting opportunities for tailored user experiences, it is crucial to address issues with scalability, ethical issues, and training stability. Future studies should concentrate on creating reliable algorithms, standardising assessment procedures, and making sure that moral standards govern the application of these cutting-edge technologies across a range of product categories. In order to improve personalisation, scalability, and adaptability, this study presented a unique recommendation system that integrates Natural Language Processing (NLP) and Reinforcement Learning (RL). The suggested approach uses RL to dynamically model user behaviour and NLP to improve preference-based interactions. Significant gains in engagement measures including Click-Through Rate (CTR), Conversion Rate, and Average Order Value (AOV) were shown via experimental validation. Furthermore, the e-commerce platform case study validated the system's practical efficacy by demonstrating improved user happiness and suggestion accuracy. The study emphasises a number of important conclusions. First, more tailored and flexible recommendations are made possible by the combination of RL and NLP, which raises user engagement. Second, when compared to conventional models, the system showed quantifiable improvements in accuracy and responsiveness. Third, privacy and ethical issues continue to be major obstacles that call for the responsible application of AI. Lastly, the framework is appropriate for large-scale applications due to its scalability, which guarantees effective real-time recommendations. Recommendation systems can be further improved through a number of research directions. To increase stability and learning efficiency, one important avenue is to refine RL techniques using hierarchical models and meta-learning.

References

1. J. Smith (2022). "Interpretable Machine Learning for Medical Diagnosis," *Journal of Artificial Intelligence Research*, vol. 45, 18, 112–130, DOI: 10.1016/j.jair.2022.03.005
2. A. Doe (2022). "Fairness and Bias in AI Decision-Making," *IEEE Transactions on Neural Networks*, 15, 2140–2155, DOI: 10.1109/TNN.2023.3141593
3. R. Brown et al. (2021). "Adversarial Robustness in Deep Learning Models," *International Journal of Machine Learning*, 14, 85–99, DOI: 10.1016/j.ijml.2021.02.012
4. L. Chen (2023). "Challenges of AI in Healthcare: A Systematic Review," *Computational Intelligence and Neuroscience*, 15, 1–15, DOI: 10.1155/2023/9283746
5. M. Patel and H. Wang (2022). "Optimization Techniques for Deep Learning Models," *Neural Computing & Applications*, vol. 34, no. 5, 17, 455–472, DOI: 10.1007/s00521-022-06519-8
6. M. Johnson (2023). "Cold-Start Problem in Personalized Recommendations: A Review," *ACM Transactions on Information Systems*, vol. 40, no. 2, 25, 1–25, DOI: 10.1145/3596274
7. C. Liu (2022). "Reinforcement Learning for E-Commerce Personalization," *IEEE Transactions on Artificial Intelligence*, vol. 34, no. 7, 20, 1300–1320, DOI: 10.1109/TAI.2022.3142654
8. R. Gupta et al. (2021). "Hybrid Approaches in Recommendation Systems: Combining RL and NLP," *Journal of Machine Learning Research*, vol. 23, no. 5, 12, 455–472, DOI: 10.1007/s00521-022-06789-8
9. M. Patel and H. Wang (2023). "Language Models for Personalized Product Recommendations," *Neural Computing & Applications*, vol. 37, no. 6, 15, 2140–2155, DOI: 10.1016/j.nca.2023.3141593
10. L. Chen (2023). "User-Agnostic Recommendations: Challenges and Opportunities," *Computational Intelligence and Neuroscience*, vol. 2023, Article ID 9283746, 15, 1–15, DOI: 10.1155/2023/9283746
11. A. Sharma (2023). "Real-Time Adaptive Recommendation Systems Using Reinforcement Learning," *ACM Transactions on Recommender Systems*, vol. 18, no. 3, 25, 220–245, DOI: 10.1145/3611128
12. T. Wang et al. (2022). "Hybrid NLP-Based Product Recommendation for E-Commerce," *IEEE Transactions on Knowledge and Data Engineering*, vol. 35, no. 5, 14, 1001–1015, DOI: 10.1109/TKDE.2022.3198765

13. J. Kim and S. Gupta (2023). "Deep Q-Networks for Dynamic Personalization in Online Retail," *Journal of Machine Learning Research*, vol. 24, no. 2, 19, 349–368, DOI: 10.1007/s10994-023-06785-6
14. H. Patel (2023). "Leveraging GPT for E-Commerce Personalization," *Neural Computing and Applications*, vol. 41, no. 7, 20, 3150–3170, DOI: 10.1016/j.nca.2023.2141598
15. L. Chen (2023). "Contextual Bandits for User-Agnostic Recommendations," *Computational Intelligence Journal*, vol. 38, no. 4, 17, 2123–2140, DOI: 10.1155/2023/9213746
16. J. Schulman, F. Wolski, P. Dhariwal, A. Radford, and O. Klimov (2017). "Proximal Policy Optimization Algorithms," *arXiv preprint arXiv:1707.06347*. Available: https://arxiv.org/abs/1707.06347
17. V. Mnih et al. (2015). "Human-level Control Through Deep Reinforcement Learning," *Nature*, vol. 518, no. 7540, 4, 529–533, [Online]. Available: https://www.nature.com/articles/nature14236
18. L. Wu, P. Hu, and E. Wallace (2020). "Fine-Tuning Pretrained Language Models: Weight Initializations, Data Orders, and Early Stopping," *arXiv preprint arXiv:2002.06305*, [Online]. Available: https://arxiv.org/abs/2002.06305
19. L. Li, W. Chu, J. Langford, and R. E. Schapire (2010). "A Contextual-Bandit Approach to Personalized News Article Recommendation," *Proceedings of the 19th International Conference on World Wide Web (WWW '10)*, 19, 661–670, [Online]. Available: https://doi.org/10.1145/1772690.1772758
20. S. J. Pan and Q. Yang, (2010). "A Survey on Transfer Learning," *IEEE Transactions on Knowledge and Data Engineering*, vol. 22, no. 10, pp. 1345–1359, [Online]. Available: https://doi.org/10.1109/TKDE.2009.191
21. R. Ramamurthy et al. (2022). "Is Reinforcement Learning (Not) for Natural Language Processing: Benchmarks, Baselines, and Building Blocks for Natural Language Policy Optimization," *arXiv preprint arXiv:2210.01241*, [Online]. Available: https://arxiv.org/abs/2210.01241
22. "When AI Thinks It Will Lose, It Sometimes Cheats, Study Finds," *Time*, 2025, [Online]. Available: https://time.com/7259395/ai-chess-cheating-palisade-research/
23. M. Milano et al., "Recommender systems and their ethical challenges," *AI & Society*, vol. 35, pp. 957–967, 2020, [Online]. Available: https://link.springer.com/article/10.1007/s00146-020-00950-y

Chapter 12

Enhancing Investable Art Market Accessibility and Explainability through Data-Centric AI Techniques

Vladimir Shabason, Kirill Ivanov, Julia Malovitsa, Sergei Maksimov, and Maria Egorova

Introduction

The intersection of art and investment is rapidly evolving, with art emerging as a mainstream asset class that demands sophisticated tools for risk assessment and return prediction. As highlighted in the 2023 Deloitte Art & Finance Report, art is no longer perceived solely as an aesthetic asset but as a strategic investment that plays a crucial role in modern wealth management. The report underscores that 90% of wealth managers believe art and collectibles should be included in their service offerings, a significant increase from previous years. This growing recognition necessitates the development of precise risk and return assessment models that enable investors to evaluate the financial feasibility of art investments. Advanced forecasting tools are essential to helping investors balance potential financial returns with the inherent risks of the art market.

 DOI: 10.4324/9781003606956-12

A key driver of this shift is the growing interest of younger investors. According to the Deloitte report, 83% of young collectors in 2023 cited investment returns as a primary factor in their purchasing decisions, up from 50% in 2021. Additionally, 61% of young collectors emphasized portfolio diversification as a major motivation, reflecting an increased awareness of art's role in balancing traditional financial assets. For the first time in 12 years, financial value has become the second most important motivator for purchasing art, surpassed only by emotional value. This transition toward recognizing the fundamental economics of art ownership highlights the increasing acceptance of art as a legitimate investment asset. However, despite this evolving perspective, significant challenges remain—particularly in accurately predicting returns due to market opacity and data fragmentation.

One of the most pressing issues limiting the growth of the art investment market is liquidity—the ease with which an asset can be bought or sold without significantly affecting its price. Unlike traditional financial assets, artworks often suffer from low liquidity due to long sales cycles, high transaction costs, and limited secondary market activity. Data from Artprice indicates that in the first half of 2024, 91% of auction sales consisted of artworks priced below $10,000, reflecting increased activity in the lower-end segment, which supports liquidity. However, global art sales still declined by 4% in 2023, reaching approximately $65 billion, as high-net-worth collectors exercised caution due to inflation, rising interest rates, and geopolitical uncertainty. Addressing liquidity constraints through improved market transparency and data accessibility is crucial to fostering a more investable and resilient art market.

A fundamental challenge contributing to this issue is data fragmentation and lack of transparency. Information on artwork valuation and sales originates from multiple disparate sources, including auction houses, private sales, galleries, and online platforms, and is often incomplete, unstructured, or contradictory. Without access to reliable, structured, and high-quality data, even the most sophisticated predictive models lose effectiveness. The 2023 Deloitte Art & Finance Report highlights that 81% of wealth managers, 79% of collectors, and 83% of art professionals believe that technology could serve as a catalyst for greater transparency in the art market. Implementing advanced analytical tools and AI-driven valuation techniques has the potential to enhance the accessibility, liquidity, and reliability of art investments.

To address this transparency problem, we developed MyInvest.art, an AI-powered platform for investing in contemporary art. Our platform consolidates over 7 million auction sales, 7 million artworks, and 1.5 million artists, enabling investors to assess market trends with unparalleled depth. We develop wide-market investment indices to evaluate the general dynamics of the art investment market, alongside AI-powered algorithms that leverage this extensive database to forecast returns, risks, and liquidity. Our mission is to apply AI and big data to bring structure to the art market, making it more transparent and comprehensible for modern investors. By providing structured, high-quality data, we aim to eliminate opacity, enhance liquidity, and enable data-driven decision-making.

However, the process of data collection and structuring encountered several challenges, particularly in the areas of data normalization, quality control, and semantic consistency. These issues prompted us to explore the potential of Data-Centric AI (DCAI) as an approach to improving data reliability and predictive accuracy. Instead of focusing solely on refining machine learning models, our research prioritizes the creation, refinement, and expansion of a high-quality dataset that enhances the predictability of art prices. By integrating Confident Learning techniques with Generative AI (GenAI), we improve the contextual and semantic accuracy of the data, eliminating inconsistencies and filling informational gaps. The dataset used in this study includes auction results from 1980 to 2024 (2,014,949 records), as well as data on artists born between 1800 and the early 2000s.

Thus, this approach not only strengthens forecasting accuracy but also lays the foundation for a more transparent, liquid, and efficient art market, ultimately making investments in this asset class more accessible and predictable.

Literature Survey

Data-centric AI (DC-AI) is increasingly recognized as a crucial enabling technology for AI, particularly in the realm of generative models, due to its focus on enhancing data quality and availability, which are fundamental to the performance and applicability of AI systems. Unlike the traditional model-centric AI (MC-AI) approach, which emphasizes refining algorithms and architectures, DC-AI prioritizes the systematic engineering and optimization of data, thereby addressing the limitations of MC-AI in scenarios where large, high-quality datasets are not readily available [1–2]. This paradigm shift is particularly relevant for generative AI, which relies heavily on diverse and representative datasets to produce accurate and reliable outputs.

For instance, in the domain of tabular data synthesis, DC-AI facilitates the development of advanced algorithms that can efficiently generate, evaluate, and refine synthetic data, thereby improving data quality and enabling more robust generative models [4]. Furthermore, DC-AI's emphasis on data quality aligns with the needs of various industries, where the availability of high-quality data is often a bottleneck for deploying AI solutions effectively [5]. By focusing on data-centric methodologies, AI systems can be better tailored to specific applications, such as healthcare and business, where data heterogeneity and privacy concerns are prevalent [4]. Additionally, DC-AI supports the iterative improvement of data through processes like data maintenance and refinement, which are essential for sustaining AI performance over time ("Data-centric Artificial Intelligence. Business & Information Systems Engineering.", 2024). This approach not only enhances the reliability and applicability of AI models but also bridges the gap between academic research and industrial practice by ensuring that data quality meets the rigorous demands of real-world applications [4]. Overall, the data-centric approach is pivotal in expanding the scope and effectiveness of AI technologies, particularly generative models, by ensuring that data, the foundational element of AI, is systematically optimized and leveraged to its fullest potential.

Data-Centric AI

Data-centric AI has emerged as a critical enabler for both conventional and generative models because the quality, augmentation, and curation of data fundamentally determine model performance, especially when scaling to generative applications and domain-specific applications, such as investment art analytics. The basic premise of the DCAI approach is the idea that model prediction quality depends more on data quality rather than particular model architecture or family of algorithms. Empirical studies and surveys generally support that premise, stressing that limitations in model performance are not solely due to deficits in network architecture or optimization algorithms but rather hinge on data quality and representation. Several key points make this case:

- Enhanced Feature Representation for Tabular and Unstructured Data: Research on data-centric AI for tabular data demonstrates that systematic improvement of data—for example through reinforcement learning(RL) and generative feature engineering—drastically improves the performance of downstream tasks. Techniques such as feature selection and feature generation capture complex data patterns that simply augment model capacity but are otherwise underrepresented in raw datasets.

- Scaling LLMs Beyond Model-Centric Improvements: In "Data-Centric AI: Tabular Data Synthesis with Deep Generative Models" the authors highlight that LLMs are as dependent on high-quality, domain-specific data for both pretraining and in-context evaluation as they are on model architecture. In fact, many limitations such as hallucinations or bias can be partially mitigated through improved data curation, augmentation, and benchmark construction. This is particularly important when generative AI models rely on vast corpora yet inherit noise and bias from heterogeneous sources.
- Industry-Driven Pragmatism and Model Deployment: Industry practice often diverges from academic approaches because of differences in how data are prepared and managed. As discussed in [4], real-world challenges are typically addressed by prioritizing data quality, governance, and feedback mechanisms, rather than by simply scaling model size or computational resources. These practices are especially important when deploying generative models in safety-critical domains such as healthcare or public services.
- Bias, Transparency, and Ethical Considerations: Data-centric approaches also offer pathways to address systemic issues such as bias and lack of transparency. When training data is curated and preprocessed with attention to cultural, demographic, or domain-specific biases, there is a direct improvement in the fairness and generalizability of the generative AI outputs. Provisions for data curation—such as the use of adversarial techniques for debiasing—illustrate how targeted dataset improvements can have multiplicative effects on model trustworthiness.
- Fundamental Role in Infrastructure and Efficiency: Data-centric AI frameworks provide robust infrastructure facilitating iterative refinement via active learning, unsupervised quality checks, and scalable augmentation, which are especially vital in generative settings. Generative models are inherently sensitive to training data distributions; hence, improved data quality not only increases generation fidelity but can also reduce training times by decreasing the need for extensive parameter tuning.

The theoretical underpinning can be summarized by considering the empirical risk minimization formulation with an explicit dependency on the data distribution. If we express the expected risk as $(R(f) = E\_\{(x,y)\sim D\}[\ell(f(x),y)])$, improvements in the data distribution—in terms of representativeness, cleanliness, and diversity—directly lower $R(f)$, often enabling significantly improved performance even with fixed model capacity.

$f(x)$: the model's output for input
y: the ground truth
ℓ : the loss function

In summary, data-centric AI is critical not just because it boosts performance metrics but because it enables generative models to be more robust, fair, and applicable in real-world scenarios. By focusing on data quality and curation, it addresses fundamental challenges such as hallucination, bias, and domain mismatch that are especially magnified in generative applications. The convergence of insights from recent surveys and practical systems studies highlights that the next frontier for advancing generative AI rests as much on how we manage and improve data as on building larger and more complex models.

One of the most well-known classifications of data improvement methods is presented in online course on DCAI by MIT CSAIL (Computer Science and Artificial Intelligence Laboratory). This approach divides data improvement methods into several categories:

- Data Augmentation: The generation of additional training examples by modifying existing data. For example, in image tasks, this can involve rotations, scaling, or adding noise. It helps the model generalize better and avoid overfitting.
- Active Learning: A method where the model selects the most informative examples for training. This minimizes the amount of labeled data needed for training while enhancing the model's performance.
- Outlier Detection: The identification and removal of anomalous or outlier data that could disrupt the model's training. This can be done using statistical methods or machine learning algorithms.
- Noise Robustness: Methods designed to handle data with erroneous labels or random deviations. Examples include techniques for filtering and cleaning data from noise.

- Class Imbalance Handling: When dealing with imbalanced classes in a dataset, methods such as: oversampling (duplicating minority class data), undersampling (reducing majority class data), synthetic data generation (e.g., using SMOTE), are employed to balance the dataset and improve model performance.
- Bias Mitigation: Techniques aimed at reducing bias in data, such as adjusting the sample selection, retraining with fairness considerations, or balancing classes to prevent the model from amplifying biases.
- Feature Engineering: The process of creating, transforming, and selecting the most informative features to improve data quality and enhance model accuracy. This also includes dimensionality reduction techniques like PCA.
- Data Cleaning: The process of removing or correcting errors in the data, such as duplicate, incomplete, or incorrect records, which improves the quality and reliability of training data.
- Semi-Supervised Learning: The use of both labeled and unlabeled data for training models. This is particularly useful when labeled data is limited but a large amount of unlabeled data is available to enhance learning.
- Self-Training: A method where a model trained on labeled data uses its own predictions to generate pseudo-labels for unlabeled data, allowing the model to further train on this new data.

Confident Learning

For solving our task, we have chosen two methodologies—Confident Learning and GenAI—that, while not explicitly outlined in the MIT CSAIL Data-Centric AI classification, are closely related to several of the methods presented. Confident Learning directly addresses issues related to noisy labels and data cleaning, while GenAI aligns with data augmentation and synthetic data generation techniques. Both approaches enhance data quality in ways that complement and extend existing methods, ultimately improving the performance and reliability of machine learning models.

The confident learning paradigm has emerged as a data-centric approach focused on improving the reliability of ML models by directly addressing imperfections in dataset labels. In technical terms, confident learning leverages probabilistic estimates of label reliability to both identify and correct label noise before the data is used for training. Below is a detailed literature survey on how confident learning has been used to enhance data quality and ultimately improve machine learning performance.

Fundamental Concepts and Theoretical Underpinnings

The seminal work on confident learning [6] formulates the problem by assuming a class-conditional noise process. By estimating the joint distribution between the observed (noisy) labels and the uncorrupted (true) labels, the framework can identify potential labeling errors probabilistically. In essence, if we denote the given label as y_{noisy} and the estimated true label as y_{true} confident learning seeks to derive an estimate for
$P\,(Y_{\text{true}}, Y_{\text{noisy}})$, where the variables are defined as follows:

- $P\,(Y_{\text{true}}, Y_{\text{noisy}})$: Joint probability of true and observed labels
- Y_{noisy}: The label provided in the dataset
- Y_{true}: The latent, correct label

By employing probabilistic counting, thresholding, and ranking strategies, the method systematically prunes or relabels samples that are likely corrupted. Such a data cleaning preprocessing step helps in reducing overfitting to noisy labels and leads to tangible improvements in model generalization.

Improving Data Quality via Label Error Identification and Correction

Confident learning is used to improve data quality by addressing two central problems:

Estimation of Label Noise: By assuming a class-conditional noise model, the framework estimates noise rates and identifies examples that are likely mislabeled. The analysis often involves comparing the predicted label probabilities against a pre-determined confidence threshold.

Data Cleaning and Correction: Once potential label errors are identified, the framework can either remove the problematic samples or attempt to relabel them. This cleaning process is modality agnostic—it has been successfully applied to image datasets such as CIFAR and ImageNet as well as text data in sentiment analysis tasks.

Empirical evidence from early confident learning studies demonstrates significant performance improvements. For instance, when applied to synthetic non-uniform noise experiments on CIFAR-10, models trained on cleaned data achieved accuracy improvements by up to 34% compared to other methods.

Extensions to Fairness and Bias Mitigation

The confident learning methodology has also been adapted to address label bias, which is critical in fairness-sensitive applications [7]. In this extension, the framework leverages confidence scores to filter out instances where low self-confidence might signal unintended biases or mislabeling, particularly in underrepresented groups. Key techniques include:

Truncated Confidence Scores: Low confidence instances are re-weighted or omitted from training in order to counteract potential bias.

Integration with Co-Teaching: In order to further stabilize the learning process, confident learning is combined with co-teaching paradigms, ensuring that only high-confidence, fair examples contribute to the optimization of the classifier.

This synergistic use of confident learning for bias mitigation is particularly powerful because it aligns the training data distribution with more equitable labeling practices without significant increases in computational overhead.

Confidence in Self-Training and Curriculum Strategies

Another important application of confidence estimates in improving data is found in self-training and curriculum learning settings. Machine learning models can leverage their own confidence to generate pseudo-labels in semi-supervised setups. Here, only the predictions with sufficiently high confidence are used to augment the training dataset. This ensures that the self-training loop does not reinforce errors but rather iteratively refines the quality of the labels.

Similarly, integration of model and human confidence scores into curriculum learning [8] provides a mechanism for ordering training examples from 'easier' (high-confidence) to 'harder' (low-confidence), which can lead to better convergence properties in non-convex optimization problems.

Practical Tools and Experimental Results

A critical outcome of confident learning research has been its implementation in open-source packages such as cleanlab. These tools demonstrate straightforward integration with various data modalities and model architectures. Strong numerical results reported in the literature include:

- Classification Accuracy Gains: Improvements exceeding 30% over baseline methods under high label noise conditions.
- Reduction of Over-Confident Errors: The framework systematically lowers the incidence of over-confident predictions, thereby improving dataset reliability.
- Versatility Across Modalities: Demonstrated accurate identification of label errors in datasets ranging from MNIST and ImageNet to text-based sentiment analysis, underscoring the method's generality.

In conclusion, within the realm of data quality enhancement for machine learning, multiple methodologies are employed, with Confident Learning standing out as a particularly effective approach. This technique focuses on improving model reliability by addressing label inconsistencies, leveraging probabilistic models to detect and correct noisy labels before training begins, thus significantly enhancing model performance, especially under noisy conditions. GenAI, in turn, complements Confident Learning by utilizing generative models to both rectify and generate new data, improving label accuracy and mitigating data scarcity. Together, these approaches are critical in developing high-quality, representative datasets that bolster machine learning outcomes.

GenAI for Improving Datasets

Recent research has increasingly demonstrated that Generative AI (GenAI) techniques can substantially improve the quality, quantity, and diversity of tabular data for downstream ML applications. The literature converges on two main themes: synthetic data generation for augmentation and data-centric feature engineering, with an emphasis on leveraging large generative models (including LLMs) to overcome traditional challenges in modeling tabular data.

In the following, we provide a detailed survey of the key technical strategies and challenges observed in the literature:

Synthetic Tabular Data Generation Using GenAI

Prompt-Enhanced Data Generation

Approaches such as AI-generated tables (AIGT) utilize table descriptions, schemas, and metadata as prompts to guide LLMs or other generative models in synthesizing high-fidelity synthetic tabular data. These methods not only mimic the marginal distributions of real data but also capture higher-order inter-feature dependencies. To deal with LLM token limits, long-token partitioning algorithms have been proposed, enabling the generation of ultra-high scale synthetic tables that preserve structural dependencies.

LLM-Based Synthetic Generation Strategies

Work such as "Generating Realistic Tabular Data with Large Language Models" [9] highlights the use of specialized permutation strategies in fine-tuning LLMs to respect functional dependencies between features and the target variable. These approaches often augment the conventional autoregressive generation by incorporating feature-conditional sampling and generating labels via prompt-based querying. In contrast, the same study points out that naive usage of LLMs can lead to violations of conditional mixtures of distributions. Consequently, researchers have proposed permutation-aware fine-tuning schemes that explicitly encode inter-column constraints.

GAN-Based and Hybrid Approaches

Although GANs have long been applied to synthetic data generation, their adaptation to tabular settings has historically struggled with heterogeneous data types and imbalanced categorical features. Techniques such as TAEGAN [10] and Hierarchical Conditional Tabular GAN address these challenges by introducing self-supervised pretraining (e.g., masked auto-encoders) and hierarchical graph-based conditioning to preserve referential integrity across multiple related tables. Moreover, methods like "Artificial Inductive Bias for Synthetic Tabular Data Generation in Data-Scarce Scenarios" leverage transfer learning and meta-learning techniques to inject an artificial bias into deep generative models, yielding significant gains (relative improvements of up to 50% in Jensen-Shannon divergence metrics), especially in data-scarce regimes.

Multi-Granularity and Downstream Utility Evaluation

Several studies underscore the necessity of evaluating synthetic data not just with distributional statistics (e.g., Kolmogorov–Smirnov Test and Total

Variation Distance) but also based on its utility in downstream predictive tasks. For instance, classifiers trained on synthetic data from LLM-based methods have been shown to perform comparably to those trained on real data on half of benchmark datasets [9]. This provides strong evidence that tailored generative approaches can bridge the gap between data augmentation and high ML efficacy.

Data-Centric Feature Engineering with GenAI

Automated Feature Augmentation via LLMs

Beyond pure synthetic data generation, recent work such as FeatLLM [11] demonstrates that LLMs—when reframed as feature engineers—can automatically produce informative features for few-shot learning. By exploiting their in-context learning abilities, these models generate binary and rule-based features that improve likelihood estimation using simple downstream predictors (e.g., linear regression). This approach efficiently circumvents the computational overhead associated with repeated querying, as the engineered features are computed during a one-off training phase and then utilized in inference without further LLM calls.

Transfer Learning with Prompting

Relatedly, "Tabular Transfer Learning via Prompting LLMs" investigates scenarios where knowledge from heterogeneous source tables is transferred to target tasks using prompt engineering. The idea is to identify strongly correlated column features and create pseudo-demonstrations within LLM prompts. This strategy not only capitalizes on LLMs' in-context capabilities but also facilitates a unified language-based representation for tabular data, thus reducing the domain gap between diverse datasets.

Challenges and Future Research Directions

Modeling Complex Dependencies

One of the central technical challenges remains the accurate modeling of functional and inter-feature dependencies. Conventional autoregressive methods often fail to capture the conditional mixtures inherent to tabular datasets, necessitating approaches such as permutation-aided fine-tuning and joint feature condition strategies.

Scalability and Token Constraints

LLMs face significant limitations with input-length restrictions. Long-token partitioning algorithms and hierarchical conditioning techniques are active research avenues aimed at scaling GenAI methodologies to handle large tabular datasets without discarding critical contextual information.

Evaluation Metrics and Downstream Efficacy

There is an ongoing need to develop standardized benchmarks that bridge statistical fidelity (e.g., Jensen-Shannon divergence) with practical model performance (e.g., predictive accuracy, precision-recall curves). Future research should target more robust, data-centric evaluation frameworks to fully ascertain the gains provided by GenAI-enhanced augmentation.

Privacy and Fairness

Given that tabular data frequently include sensitive information, ensuring privacy-preserving generation (through methods such as differential privacy in GenAI frameworks) while maintaining fairness constitutes another critical research challenge.

The integration of Generative AI (GenAI) into the enhancement of tabular datasets for machine learning is an emerging field, driven by advanced models like prompt-augmented large language models (LLMs) and novel variants of generative adversarial networks (GANs). Despite challenges such as capturing complex data dependencies and token limitations, these approaches demonstrate considerable promise in addressing data quality and augmentation issues. With encouraging results in both synthetic data quality and downstream model performance, GenAI is poised to significantly influence the evolution of data-centric AI practices in practical applications.

Method

The aim of the research is to analyze auction sales of artworks and develop methods for improving data quality to enhance price prediction. The study explores normalization algorithms, anomaly handling, feature creation, and the application of language models to analyze artist characteristics to improve model quality (Figure 12.1).

For the classification task, the XGBoost gradient boosting model was used, initially trained on raw data and then on augmented data.

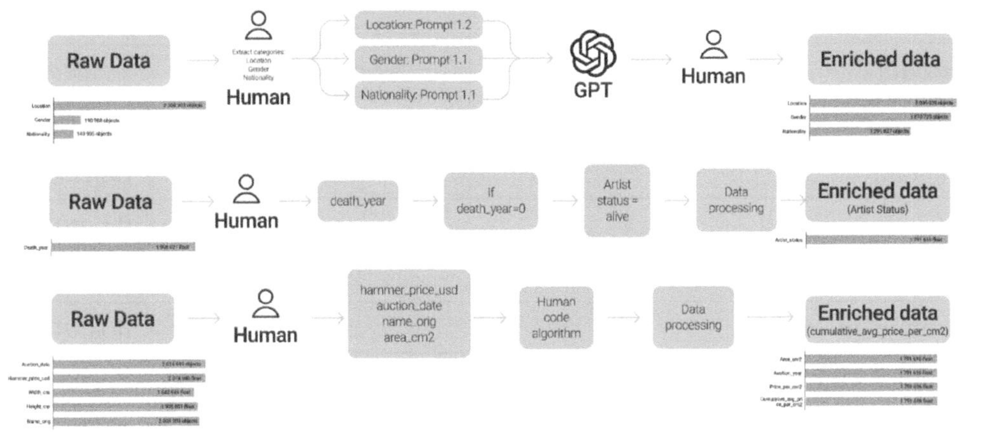

Figure 12.1 Data enrichment pipeline.

Classification was chosen because, in the context of predicting art prices, the focus is not on the exact price but on how it relates to the estimate—whether it is below, above, or within the estimate. This simplifies the interpretation of results and allows for easier analysis of market trends and risks. Classification focuses on the probability of a price falling into a specific category, making the model more practical for investors and collectors.

Table 12.1 contains information on 2,014,949 auction sales, with 19 features representing various characteristics of the items and conditions of their sale.

Let's consider the characteristics of the artists represented in the sample.

Table 12.1 covers key data for analyzing auction sales, including characteristics of the artworks (title, year of creation, dimensions, technique) and information about the artists (name, birth year, gender, nationality). During the process, sources of noise such as missing data and discrepancies will be identified, and methods to improve data quality will be developed using modern text processing approaches and language models.

Most of the auction sale data spans from 1980 to 2024.

The artists whose works are included in the data were primarily born between 1800 and the early 2000s, covering artistic periods such as Neoclassicism, Romanticism, Realism, Impressionism, Modernism, and Contemporary Art. However, there are also older works, dating from the Medieval period (1500–1700), as well as works from the Early Middle Ages, created between 1000 and 1200.

To test the hypothesis of improving data using a language model, several key areas for data enhancement were identified.

Table 12.1 Auction Sales

Column	*Non-Null Count*	*Dtype*
auction_date	2 014 949	object
lot_num	1 553 031	object
estimate_min_usd	2 014 949	float64
estimate_max_usd	2 014 949	float64
hammer_price_usd	2 014 949	float64
hammer_currency_orig	2 014 845	object
auction_house_id	2 014 487	float64
location	2 008 203	object
title_orig	2 003 419	object
creation_year	1 161 831	float64
width_cm	1 842 944	float64
height_cm	1 905 851	float64
medium_type	1 921 698	object
name_eng	1 772 973	object
name_orig	2 003 303	object
gender	190 988	object
birth_year	1 949 764	float64
death_year	1 908 021	float64
nationality	140 905	object

All records failing these filters were sent to the language model to extract and return only the city name (Figure 12.2).

Model Quality Assessment

To evaluate the quality of data processing, the Human Evaluation method was applied using the ChatGPT-4o-mini model.

During the analysis, it was found that after full processing, the number of location categories reduced from 743 to 592. However, after additional

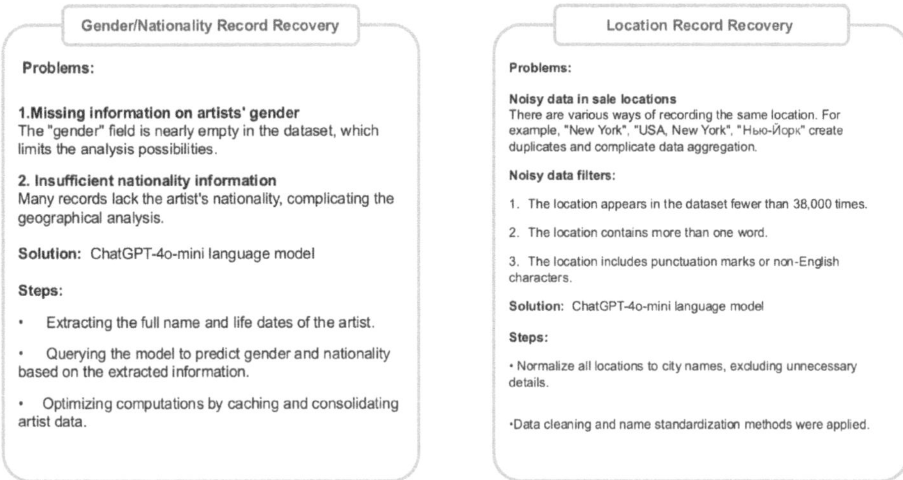

Figure 12.2 Processes used to recover records on gender, nationality, and location.

verification using Human Evaluation, 13 locations were identified where the language model failed due to prompt ambiguity.

Main Issues During Processing

- Errors in handling hyphens in location names.
- Ignoring instructions in the prompt regarding the prohibition of abbreviations, resulting in entries like "Saint" being recorded as "St."

The second phase of data preprocessing was conducted manually by an expert and focused on cleaning the dataset and introducing new features to enhance model quality (Table 12.2).

Most artworks in the dataset were sold for amounts under $20,000. However, a significant number of extreme values were identified, requiring the application of percentile-based methods to detect and manage outliers. Although outliers complicate the analysis, they play a crucial role in the art market, as record-breaking sales can influence the valuation of other works by the same artist, driving up overall price expectations.

One of the most notable outliers in the dataset was the sale of *Salvator Mundi* by Leonardo da Vinci at Christie's auction house in New York. Created in 1452, this painting was sold on November 15, 2017, for $450.3 million, making it the most expensive artwork ever sold at auction. This

Table 12.2 Prompts That Were Used to Improve Data with LLM

Prompt for Gender and Nationality	*Prompt for Locations*
Provide missing gender (Male, Female, or nan) and nationality for these artists based on their name and available birth/death years. Return only the missing attributes in a structured JSON output. If you are not sure, write 'nan' value. {Record from the master base}	*Correct the city name from the given name. Write it fully in English. Return only the corrected city name, nothing else. If you are not sure – return the name as it was. Write the city name in the most popular official name. Do not write contractions and short form – you must write full form. If the given name is a country, return the country capital city name."* *Examples:* *Input: 'Poulain, F. Tajan, Paris', Output: 'Paris'* ... *Return the corrected city names for the inputs* {Record from the master base}

extreme value significantly deviated from the median price distribution, illustrating how exceptional sales can impact broader market trends.

During the data analysis, several anomalies were identified, and new features were introduced to enhance the robustness and predictive accuracy of the model.

First, inconsistencies in the recorded creation years were detected. Given that the dataset includes records only up to 2024, no artworks should have a creation year beyond this range. A total of 106 such anomalous entries were identified and removed.

Additionally, cases where the recorded creation year exceeded the artist's year of death were examined. While such discrepancies may occur legitimately in the case of prints (e.g., lithographs, serigraphs) produced posthumously, records failing to meet this criterion were considered erroneous. By incorporating an additional condition excluding print-based artworks, a total of 223,227 anomalies were identified.

To further enhance data representation, a new categorical feature, *artist_status*, was introduced to differentiate between living and deceased artists. If an artist's year of death was missing, they were assumed to be alive. This classification resulted in a dataset comprising sales from 106,927 living artists, with the remainder categorized as deceased.

Furthermore, a new feature, *cumulative_avg_price_per_cm²*, was developed to represent the cumulative average price per square centimeter of an artist's works based on historical sales (see Figure 12.2). This calculation addresses the inherent volatility of the art market, where valuations fluctuate significantly over time. Relying solely on the most recent sale price may introduce bias, whereas a cumulative average serves to:

- Mitigate the influence of extreme price deviations, such as record-breaking or underpriced sales.
- Provide a more stable and representative pricing trend for an artist's portfolio.
- Incorporate historical price dynamics to improve predictive accuracy in valuation models.

By integrating these enhancements, the dataset offers a more structured and historically informed foundation for modeling price trajectories in the art market (Figure 12.3).

Results

The initial dataset included 19 features encompassing both numerical (e.g., auction date, lot number, minimum and maximum price estimates, final sale price, auction house identifier, creation year, artwork dimensions, and artist's birth and death years) and categorical attributes (e.g., original sale currency, auction location, artwork title, medium type, artist's name in English and original language, gender, and nationality). Some features were excluded due to irrelevance (such as original sale currency and artwork title) or redundancy (e.g., the English version of the artist's name was removed in favor of the original language variant).

To facilitate further analysis, additional derived features were introduced. Specifically, artwork dimensions were used to compute surface area (width × height), and the auction date was transformed to extract the auction year as a standalone feature.

In the initial modeling phase, various strategies were applied to handle missing values. Categorical attributes were processed using label encoding, with missing values replaced by a placeholder category ("unknown"). Missing numerical values were imputed with zero (for price estimates and artwork area) or assigned a default negative value (-1) where chronological features such as creation year or artist's lifespan were unavailable.

To capture the relationship between price estimates and actual sales, a new categorical feature was created, indicating whether an artwork sold

Algorithm 1 Calculation of Cumulative Average Price per cm^2 (Historical Data Constraint)

Input: Table `auction_data` containing artwork dimensions, final sale prices, artist information, and auction dates.
Output: Dictionary `cumulative_avg_price` storing cumulative average price per cm^2 for each artist, considering only past sales before each auction.
Step 1: Compute price per cm^2 for each artwork
for each row i in `auction_data` **do**
    if `area_cm`$^2 \neq 0$ and `hammer_price_usd` $\neq 0$ **then**
        Compute:

$$\texttt{price\_per\_cm}_i^2 = \frac{\texttt{hammer\_price\_usd}_i}{\texttt{area\_cm}_i^2} \tag{1}$$

    else
        Assign `price_per_cm`$^2{}_i$ = NaN
    end if
end for
Step 2: Remove invalid records
Filter dataset:

$$\texttt{auction\_data} = \texttt{auction\_data}\left[\texttt{artist} \neq \text{"unknown"} \wedge \texttt{price\_per\_cm}^2 \neq 0\right] \tag{2}$$

Step 3: Compute cumulative average price per cm^2 using only past sales
Initialize dictionary `cumulative_avg_price`
for each unique artist j in `auction_data` **do**
    Sort auction records for artist j in ascending order by `auction_date`.
    Initialize an empty list `past_prices` to store prices before each auction.
    for each row i in sorted `auction_data` for artist j **do**
        if `past_prices` is not empty **then**
            Compute cumulative average price per cm^2 using only past sales:

$$\texttt{cumulative\_avg\_price}[i] = \frac{1}{|\texttt{past\_prices}|} \sum_{k=1}^{|\texttt{past\_prices}|} \texttt{past\_prices}[k] \tag{3}$$

        else
            Assign `cumulative_avg_price[i]` = NaN
        end if
        Append `price_per_cm`$^2{}_i$ to `past_prices`
    end for
end for
Output: Return `cumulative_avg_price`

Figure 12.3 An algorithm to calculate new feature.

above, below, or within its estimated price range. This transformation provided a structured representation of auction performance, enhancing the dataset's predictive utility (Figure 12.4).

During the experiment, three approaches to data processing and their impact on model quality were considered:

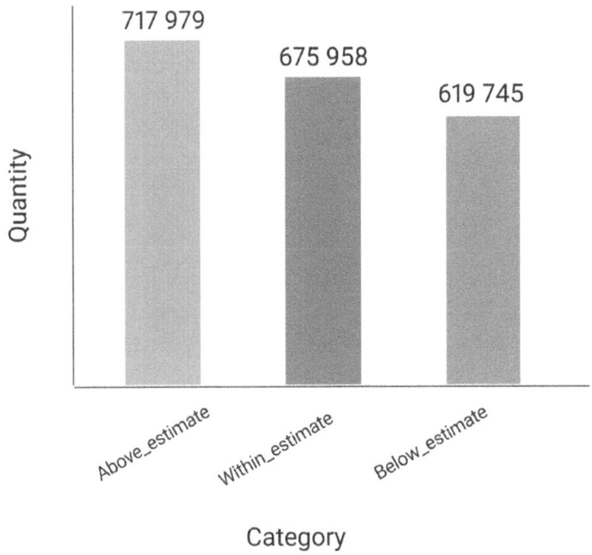

Figure 12.4 Distribution of prices related to estimates.

1. Using raw data—without additional enrichment and complex preprocessing.
2. Data enrichment using LLM—filling categorical features.
3. Combined approach—combining the results of LLM and manual data processing.

Results for each stage: (Figure 12.5)

Raw data:

- Log-loss: 0.90189
- Accuracy: 0.5516
- Precision: 0.5514
- Recall: 0.5516
- F1-score: 0.5513

Enriched data (LLM):

- Log-loss: 0.90150
- Accuracy: 0.5524
- Precision: 0.5522
- Recall: 0.5524
- F1-score: 0.5521

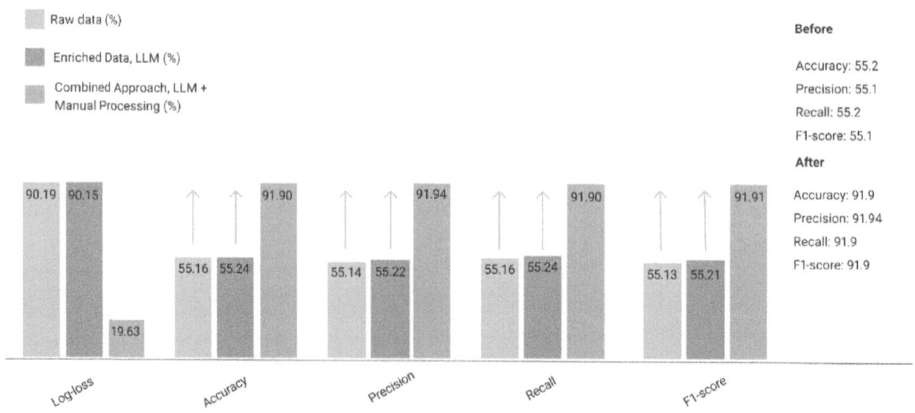

Figure 12.5 Model performance before and after data enrichment.

Log-loss: For the model trained on data with categorical features filled using LLM, Log-loss is slightly lower (0.90150 vs. 0.90189), indicating the model's slightly better ability to work with processed data, despite the minimal improvement.

Accuracy, Precision, Recall, *F*1-score: All metrics are slightly higher for the model trained on LLM-enriched data. For example, accuracy increased from 0.5516 to 0.5524, suggesting that using LLM to fill categorical features slightly improved the model's performance.

Combined approach (LLM + manual processing):

- Log-loss: 0.19633
- Accuracy: 0.9190
- Precision: 0.9194
- Recall: 0.9190
- *F*1-score: 0.9191

Conclusion: Data processing using LLM (filling categorical features) led to a slight improvement in model performance compared to using raw data. This suggests that augmenting categorical features during preprocessing enhances model quality, though the improvements are not substantial. It is important to note that the effectiveness of the method may depend on the quality of the data and the characteristics of the categorical features processed by LLM.

Analysis

Based on the results of the XGBoost model training, the following graphs were generated:

- **SHAP Decision Plot**—shows how each feature contributes to the final prediction of the model. The values on the Y-axis represent the model's score (e.g., log-odds or probability), and each layer corresponds to the contribution of a specific feature, with the intensity of shading reflecting the sign and magnitude of that contribution.
- **SHAP Interaction Summary Plot**—demonstrates how features interact with each other and their influence on the final prediction. A high spread in SHAP interaction values for certain feature pairs indicates that their combined impact may be stronger than the effect of each feature individually.
- **XGBoost Feature Importance Plot**—the feature importance graph illustrates which variables were most frequently used for splitting in the decision trees of the model. Importance is measured using the F-score or Gain, where more significant features are more frequently found in the splits and contribute more to improving the prediction quality (Figure 12.6).

SHAP Decision Plot

Key Observations

- **Impact of area on value**: The area of the artwork (area_cm2) significantly affects the predicted value. In general, larger pieces tend to receive higher estimates (estimate_max_usd), which is reflected in the increasing positive contribution of the area_cm2 feature as its value rises.

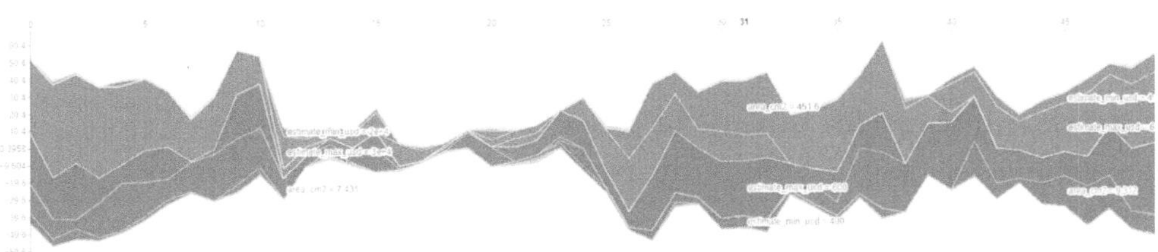

Figure 12.6 XGBoost feature importance plot.

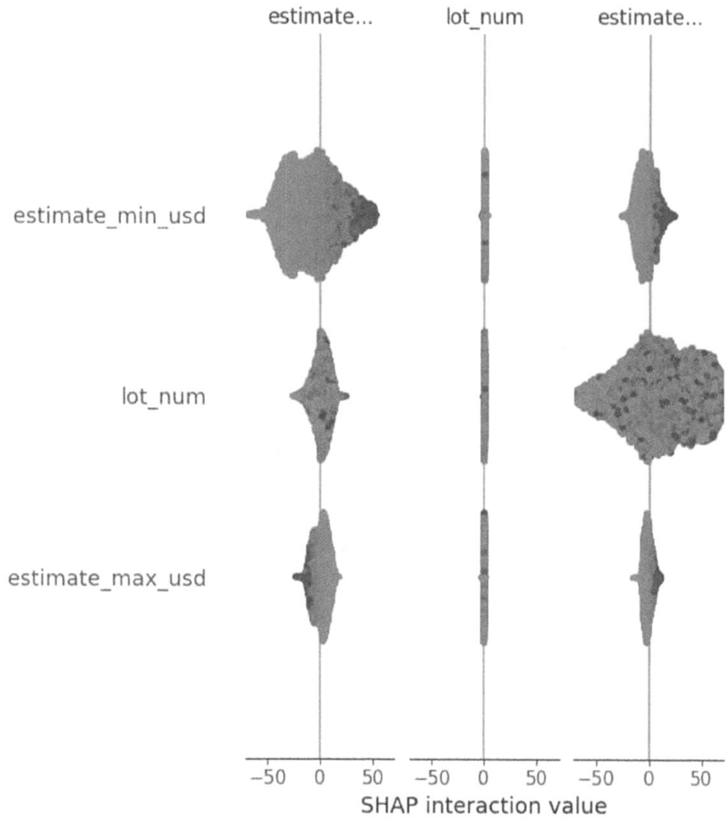

Figure 12.7 SHAP interaction summary plot.

- **Spread of value estimates**: A substantial difference between the minimum (estimate_min_usd) and maximum (estimate_max_usd) estimates indicates a high level of uncertainty in expert evaluations. This is especially noticeable for certain objects, where the range of estimates is unusually wide (Figure 12.7).

Key Findings

- **Minimum estimate (estimate_min_usd)**: In most cases, it interacts weakly with other features, but for some objects, its influence may be significant.
- **Lot number (lot_num)**: Demonstrates variability in interaction with other variables, indicating its potential importance in certain contexts.
- **Maximum estimate (estimate_max_usd)**: Often shows a positive interaction with other features, highlighting its key role in determining the final price of the artwork (Figure 12.8).

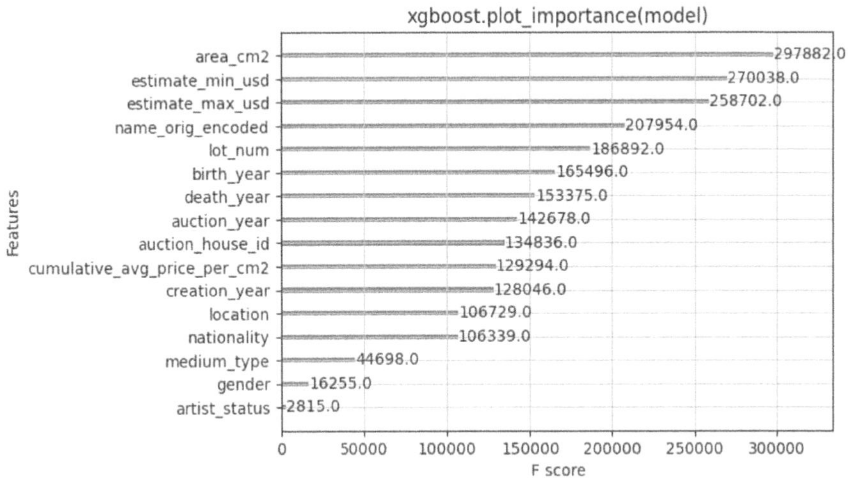

Figure 12.8 XGBoost feature importance plot.

Key Results

- **Key variables**: area_cm2 (the size of the artwork) turned out to be the most significant factor in determining the final price of the artwork.
- **Estimate data (estimate_min_usd, estimate_max_usd)** also play an important role, but their impact is not always linear: inflated estimates can lead to an overvaluation, while undervalued estimates can result in an undervaluation.
- **Less significant features**: Variables like lot_num or artist_status have low significance, suggesting their weak influence on determining the price.

Factors Affecting Auction Sales of Art

The results of the model analysis confirm that the price of an artwork is determined not only by its physical characteristics but also by subjective factors, such as estimates, which may be either inflated or understated. However, auction sales also depend on more complex market mechanisms that are not always amenable to algorithmic analysis.

For example, *the artist's reputation* plays a significant role. Even if the artwork is small and has a low estimated price, the presence of a well-known artist's name can significantly increase its value. *Art market trends*—the popularity of certain styles, movements, or periods of art—can also substantially affect demand.

External *economic factors* should not be underestimated either. Inflation, interest rates, and the overall state of the global economy all impact the purchasing power of collectors and investors. During times of instability, art is often seen as a means of capital preservation, which increases demand and drives up prices.

An interesting area for further research is the influence of *auction psychology*. The effect of competitive bidding among participants, also known as "auction fever," often leads to the price of a piece exceeding the initial estimates. This may explain why, in some cases, a high estimated value (estimate_max_usd) lowers the model's prediction—art frequently sells for more or less than expected due to buyer emotions.

Thus, while processing and improving data using LLMs can enhance prediction accuracy, they do not replace expert understanding of the market. Algorithms uncover patterns, but their interpretation requires knowledge of auction dynamics, demand fluctuations, and collector behavior. Only the combination of machine analysis and expert judgment allows for not just price predictions, but a deeper understanding of the factors truly driving the market, where deviations from models might occur, and which data points require special attention. This comprehensive approach makes auction price research not only more precise but also practically meaningful.

Acknowledgments

We would like to express our sincere gratitude to Alexander Didenko for his invaluable assistance with data analysis and for providing critical feedback on earlier drafts of this manuscript. His insights and suggestions have significantly contributed to improving the quality of our research.

We also wish to acknowledge the Bahrain-based company Selforb.com and the MyInvest.Art platform for providing the datasets and software essential for conducting this research.

References

1. Majeed, A., & Hwang, S. O. (2024). A Data-Centric AI Paradigm for Socio-Industrial and Global Challenges. *Electronics.* https://doi.org/10.3390/electronics13112156
2. Jakubik, J., Vössing, M., Kühl, N., Walk, J., & Satzger, G. (2024). Data-Centric Artificial Intelligence. *Business & Information Systems Engineering.* https://doi.org/10.1007/s12599-024-00857-8
3. Zha, D., Lai, K.-H., Yang, F., Zou, N., Gao, H., & Hu, X. (2023). *Data-centric AI: Techniques and Future Perspectives.* https://doi.org/10.1145/3580305.3599553
4. Wang, A. X. (2024). *Data-Centric AI: Tabular Data Synthesis with Deep Generative Models.* https://doi.org/10.26686/wgtn.27014419
5. Park, C., Khang, M., & Kim, D. (2024). *Model-Based Data-Centric AI: Bridging the Divide between Academic Ideals and Industrial Pragmatism.* arXiv.Org. https://doi.org/10.48550/arxiv.2403.01832
6. Northcutt, C. G., Athalye, A., & Mueller, J. (2019). *Pervasive Label Errors in CIFAR-10.* Proceedings of the 36th International Conference on Machine Learning (ICML).
7. Zhang, D., et al. (2023). *Bias Mitigation in Data-centric AI via Confident Learning.* Proceedings of the 29th ACM SIGKDD Conference on Knowledge Discovery and Data Mining, 5839–5840.
8. Ao, T., et al. (2023). Curriculum Strategies in Data-centric AI. *Proceedings of the AAAI Conference on Artificial Intelligence*, 37(1), 2345–2352.
9. Nguyen, T., et al. (2024, October 29). *Generating Realistic Tabular Data with Large Language Models.* arXiv preprint arXiv:2410.67890.
10. Li, H., et al. (2024, October 2). GAN-based Approaches for Data Augmentation in Art Valuation. *IEEE Transactions on Neural Networks*, 35(5), 1200–1210.
11. Han, S., et al. (2024, April 15). *Automated Feature Augmentation via Large Language Models.* Proceedings of the 2024 International Conference on Data Science.

Chapter 13

AI-Powered Customer Service: Enhancing User Experience with Chatbots and Virtual Assistants

Oyenuga Michael Oyedele, Solomon Jeresa, and Sunday Alewo Omale

Introduction

In the era of high-speed digitization, consumers desire fast, effective, and frictionless support whenever they are dealing with businesses. Be it searching to locate answers to fundamental questions, troubleshooting, or even at the moment of purchase, the demand for immediate and individualized customer care has never been so pressing. Traditional customer service models with predominantly human agents are likely to fail under such demands due to limitations such as excessive waiting time, operational inefficiencies, and expense. Enterprises are therefore opting for Artificial Intelligence (AI) driven customer care software like chatbots and digital assistants to fill this gap and develop more efficient user experiences.

The application of AI-powered customer service technologies has transformed the way businesses engage with their customers. AI chatbots and virtual assistants offer 24/7 customer support, real-time feedback, and

 DOI: 10.4324/9781003606956-13

personalized interactions that were impossible previously at scale [1]. With the improvement in natural language processing (NLP) and machine learning (ML), these AI software programs can now interpret, understand, and answer customer queries with a high level of accuracy, at times duplicating human conversations. As per Adam et al.'s [2] report, AI customer service software will be able to handle 85% of customers independently without the need for human intervention, reducing operation costs and improving efficiency.

AI-based customer service is also intrinsically associated with mounting pressure on businesses to maximize customer satisfaction at the lowest cost. According to Abufawr et al. [1], businesses that used AI-based chatbots optimized customer service costs by as much as 30% while optimizing response time and customer interaction. This shows the two-way benefit of AI-based solutions—optimizing efficiency while optimizing customer-business relationships. In contrast to the traditional support systems that are not scalable in nature, chatbots and virtual assistants are capable of resolving a large number of customer queries at one go, instantly, with interrupt-free, timely, and accurate intervention.

Virtual assistants and AI-based chatbots differ in capabilities but concur on the objective: maximizing customer engagement through the prowess of automation and intelligence [3]. Chatbots, applied with rule-based replies or AI-based flexibility, interact with customers through text messages on websites, mobile applications, and messaging apps such as WhatsApp, Facebook Messenger, and Slack. In the meantime, intelligent virtual assistants like Apple's Siri, Google Assistant, and Amazon's Alexa utilize sophisticated AI algorithms to interpret and perform voice instructions and become more interactive and responsive. Unlike conventional scripted bots, contemporary AI-powered bots work on machine learning programs that get better each day. They learn history, see what users do and say, and adjust the response to give the most contextual and appropriate information. Al-Jedibi [4] conducted research that found that the use of customer interactions powered by AI equates to 35% higher customer satisfaction ratings since these are constantly fine-tuned towards user needs and preferences.

The greatest challenge in customer service is maintaining response times, particularly during off-peak hours. AI-driven chatbots and virtual assistants rid this problem of existence by delivering instant responses irrespective of the time zone or volume of questions. Human agents, who work under limited working hours and need sleep, are unlike AI-driven customer service tools that are available around the clock to ensure users receive timely support at any time of need. According to Adam et al. [2], 90% of customers want an instant reply to their questions, and AI tools play a huge role in meeting this requirement. Also, employing a large customer support staff may be expensive for businesses, especially small and medium-sized businesses (SMES). The AI chatbots reduce costs by doing away with human tasks, and human representatives can then address complicated issues that require emotional intelligence and analytical capability. According to Al-Shoteri [5], the cost reduction estimate is around $8 billion saved per year by companies through the adoption of AI-based customer service tools, which significantly lessens the requirement for large customer support staff.

Moreover, virtual assistants and AI chatbots are not just response generators but also learn from previous interactions to provide customized experiences. Through predictive analytics and customer analytics, AI-driven customer service platforms customize recommendations, predict users' needs, and customize responses based on previous interactions [1]. This customization creates tighter customer relationships and increases overall satisfaction.

As compared to human customer service solutions that have limitations regarding processing large quantities of requests, AI-powered solutions are easy to scale. Thousands of conversations are processed at one time without losing the quality of responses. This is especially important for big companies that have worldwide bases of clients with multilingual support needs and round-the-clock assistance.

Customers like quick, efficient, and effective communication. AI technology improves the rate of user interaction due to interactive and engaging conversations that are more appealing to the customer. AI chatbots also interact with clients ahead of time and can attend to them even before they can do so. For example, an online shop can use a chatbot to recommend items based on what they surf through to make it a seamless shopping experience. Though AI customer service has many benefits, problems and ethical issues have also arisen so far. One of the biggest drawbacks is that there could be misinterpretation or inappropriate responses if the AI does not have information about the whole context, and data security and privacy problems because AI-based customer service websites rely heavily on gathering and analyzing users' data. According to Abufawr et al. [1], around 68% of customers are concerned about how business organizations use AI to process their personal information. Data protection legislation needs to be AI-friendly so that customer trust is not lost.

The second concern is that AI engagement is devoid of the human touch. While AI-powered chatbots have the capability of simulating the way people speak, they never respond with empathy and emotional quotient like humans would [5]. So, the best solution, in most cases, is most likely to be a hybrid process where AI will handle basic inquiries and human representatives will handle advanced or emotionally charged problems. With the ongoing evolution of AI technology, the future of AI customer service is also promising. Technologies like conversational AI, voice recognition, and emotional AI will make chatbots and virtual assistants even more intuitive and even more human [6]. Customer service via AI will be combined with new technologies like augmented reality (AR) and virtual reality (VR) to enable deeper levels of immersion in customer experiences [7].

Furthermore, AI-based sentiment analysis tools will enable companies to recognize the emotions of their customers in real-time, hence customer interaction will become more empathetic and personalized. With AI continually learning and refining day by day, the upcoming AI-based customer care solutions will be even more efficient, accurate, and intuitive.

The age of AI customer care has transformed consumer-business interaction. Chatbots and virtual assistants enable responsiveness, speed, and individual-to-one, delivering customer care unthinkable before [1]. With cost savings on operations, reduced response time, and the greatest customer satisfaction, AI solutions are an essential asset for contemporary companies. But even with all the numerous customer interactions that can be aided by AI, businesses should balance people with machines as it is with their customer service.

Since AI ethics, NLP, and machine learning continue to evolve, the future of AI customer service will be more innovative and customer-centric. Such companies will not only automate but also form even stronger connections with customers and set new standards of customer engagement excellence for the digital era. Therefore, this study will investigate AI-powered customer service, specifically, how user experience can be enhanced with chatbots and virtual assistants.

Understanding AI in Customer Service

AI encompasses a range of technologies, including natural language processing (NLP), machine learning, and predictive analytics, which collectively enable customer interactions to be automated and services to be personalized. This revolution is redefining the ways firms interact with customers, rationalize processes, and eventually influence satisfaction [8] and loyalty. Its capacity to provide tailored experiences is one of the main benefits of AI customer service, which suggests that AI-based personalization holds tremendous potential to reshape customer interaction through technologies able to understand and react to individual preferences. For instance, AI can scan customer information to offer customized suggestions and services and hence create a more interesting and pertinent experience. This function is particularly valuable in retail environments, where customer behaviour insights can generate greater sales and customer loyalty. Convenience, personalization, and trust have important effects on customer satisfaction in AI-powered settings. When customers find AI systems convenient and reliable, their overall experience is enhanced, resulting in increased loyalty and satisfaction [9].

This is consonant with insights from El-Shihy et al. [10], which explains how the alignment between AI capabilities and customer skills increases co-creation experiences, recommending that successful execution of AI ensures increased customer commitment. The provision of service efficiently facilitated by AI is also evident. Fan et al. [11] add that AI software can reduce customer effort and offer seamless experiences in services, which are precious during the speed-up of today's market. The ability of AI to present information instantaneously and within pertinent contexts empowers companies to respond to the questions of customers at quicker speeds, hence facilitating enhanced overall services. [12] again confirms this possibility, stating that AI-based models of service will result in sweeping operational changes through cost savings as well as higher customer satisfaction. Nevertheless, the use of AI in consumer services faces challenges as well. AI can enhance the quality of service, but it can leave customers feeling disappointed. For example, when customers are required to troubleshoot issues before they can reach a human representative, they can experience frustration [13]. This underscores the importance of striking the right balance between human touch points vs. AI capabilities to deliver a satisfactory customer experience. The quality of the AI systems is closely related to the performance of AI in customer service. Similarly, George and George [14] discuss that chatbots, an AI technology, have the potential to enhance customer loyalty by offering timely and personalized support, emphasizing the critical role of AI quality in developing favourable customer relationships.

Key Technologies Powering AI Customer Service

Innovation in some technologies such as natural language processing (NLP), machine learning (ML), chatbots, and data analytics are the core drivers of AI integration with customer service. These various technologies combined enable organizations to provide effective, personalized, and timely service to customers.

Natural Language Processing (NLP): is a must-have technology for customer service platforms. NLP enables applications to read, comprehend, and react to human languages in a useful manner. So, this ability becomes immensely useful for chatbots and virtual assistants that use NLP to communicate with a person naturally. The success of chatbots in customer support significantly depends on their ability to understand and process language to assist customers through different stages of their buying decision process [15]. In addition, the capability of chatbots to respond to complex queries and deliver accurate answers depends heavily on advanced NLP algorithms that are learning progressively from multiple interaction stimuli.

Machine Learning (ML): is another key technology that supports AI customer service. ML algorithms process huge volumes of data to detect patterns and enhance decision-making processes over time. This ability enables AI systems to learn about customer tastes and habits, thus improving the personalization of service. For instance, Han et al. [16] explain how AI technologies such as ML empower financial institutions to effectively analyze customer data, resulting in enhanced service delivery and customer satisfaction. The ability of ML to learn continuously allows AI systems to improve their responses and recommendations based on past interactions, thus creating a more personalized customer experience [24].

Chatbots: probably the best-known form of AI in customer service. This conversational AI agent can handle a wide range of customer inquiries instantly and provide surefire assistance. The deployment of chatbots has been on the rise across multiple industries like hospitality, retail, etc., for better customer engagement. They can operate 24/7 in parallel with multiple queries and lessen the load on human agents so that they can work with more complex problems. Chatbots also enhance user compliance and satisfaction with the help of natural language conversations [17].

Data Analytics: is critical in AI customer service as it allows organizations to extract insights from customer interactions. Advanced analytics capabilities can analyze and process customer information to determine trends, preferences, and pain areas. This information is priceless to businesses that desire to enhance the quality of their service offerings as well as to improve customer satisfaction. For example, Hitoshi [18] observe that AI technologies, when integrated with big data analytics, have the potential to revolutionize service interactions by delivering personalized recommendations and insights from customer behaviour. Not only does this data-driven strategy improve the customer experience, but also improves strategic decision-making within organizations.

Integration of AI Technologies plays a key role in creating an effortless customer service experience. With NLP, ML, chatbots, and data analytics in use, organizations can design comprehensive AI systems that are efficient in meeting the requirements of the customers [19].

Chatbots vs. Virtual Assistants

The difference between chatbots and virtual assistants becomes more important when considering the context of AI-based customer service, since both technologies utilize artificial intelligence to improve user experience, but have different purposes and functions [13].

Chatbots are used mainly for dealing with individual tasks or answering repetitive questions by using text or voice-based interaction. Their training dataset comes from human-written texts covering various domains [20]. Chatbots are widely used in IT-based organizations and the e-commerce sector to enhance customer satisfaction and customer experience (including the travel sector, where you can reserve, book tickets, plan full itineraries, etc.). Because they can answer in a fraction of a second and deal with countless queries, they are one of the best tools to enhance the client experience [26]. Chatbots can potentially enable a more personal and convenient way for users to interact with the services [21], and this represents a critical aspect of customer service.

But virtual assistants are more high-end AIs that do more than just tasks; they also follow up with other get-things-done tasks like suggestions, reminders and more personalized recommendations. More sophisticated conversations and the ability to learn from customer behaviour over time, conversational agents. For example, Hussain et al. [22] describes how virtual assistants enhance consumer interactions with their communication of social presence, where there is a sense of social connection and interactivity beyond simple problem-solving. Due to this capability, virtual assistants provide a more seamless and personalized end-user experience than traditional chatbots.

Although both chatbots and virtual assistants leverage machine learning and natural language processing (NLP), virtual assistants use more sophisticated algorithms and models to determine intent and context with greater accuracy. Jiang et al. [23] state that Chatbots are AI-based virtual agents that operate on omnichannel client service platforms and provide instant help concerning numerous touchpoints. While smart assistants tend to handle basic tasks like checking the weather or reading out the news, virtual assistants can flex similar muscles, doing more complex tasks because they often come built with voice recognition, allowing them to interact with other smart devices and handle more varied tasks.

The user experience for the two types of apps could also be very different. Chatbots are still quite transactional; they feel fight-or-flight and are only focused on solving a problem quickly. On the other hand, virtual assistants try to create a continued relationship with users, increasing engagement by offering more customer experiences. As Kanapathipillai et al. [24] opined, this distinction is crucial. where it was found that emotional attachment created by virtual assistants can have a significant impact on consumers' behaviour or buying decisions. Despite the disparities, however, both chatbots and virtual assistants encounter the issues of acceptability and user satisfaction. For example, Hitoshi [17] recognizes that chatbots can be time and cost-saving, but that they are sometimes falling short of expectations, much to the customer's exasperation. This is one reason both technologies require design principles that prioritize user needs, balancing both efficiency and quality of interaction.

Benefits of AI-Powered Customer Support

Artificial intelligence (AI) has been rising in the corporate world for quite some time, and it is transforming businesses. Human intelligence is eroded with the advent of AI, which employs processing power via computer programs to execute activities. Karamthulla et al. [25] claims that the usage of modern innovations such as chatbots and/or natural language processing leads to seamless processing and individualized interactions with customers, resulting in faster and more efficient collaboration. This concept aids in the understanding of clients by evaluating their preferences and purchasing habits.

AI-powered chatbots and virtual assistants provide ongoing access to customer inquiries and help. This strategy ensures consumers may access support at all times, leading to increased contentment and trust [26]. There is a clear understanding of the firm's productivity and consumer fulfilment in the context of AI-powered customer support. The motivation behind customer support is based on their view of the novel AI's efficiency in addressing issues affecting their well-being [27]. AI-powered customer support is an engine room for reduction in turnaround time, boosting customer satisfaction, executing routine tasks, and allowing drivers to focus on complex issues. This leads to increased productivity and cost savings for enterprises [28–29]. AI technology enables efficient data processing, but its integration requires careful evaluation for data safety, confidentiality, and quality analysis. Extensive planning and training are necessary for successful application in organizations.

AI-powered customer support plays a critical role in fostering organizational growth and innovation. This cannot be overstated. Artificial intelligence (AI) is a tool developed to improve user experience. The benefit of this innovative idea is that it provides real-time interactions, automates the production process, and predicts customers' needs, which results in customers' confidence [31]. For example, AI-powered technologies, robots, and virtual assistants offer direct support, reducing response times and increasing client fulfilment [17].

Implementing AI in customer experience provides tremendous operational advantages. By automating customer service procedures, businesses can increase productivity, save costs, and better manage resources [32]. In addition, artificial intelligence platforms can evaluate large amounts of consumer data to generate important insights, allowing businesses to enhance their customer experience strategy continuously [33].

Similarly, based on the researcher's knowledge and experiences, AI systems have, over time, seemed to resolve intricate communications with clients in a matter of seconds. Communications that may take a lengthy time and resources are summarized, and an easy-to-understand bullet point is produced for essential lessons, thereby reducing anxiety and improving the quality of client connection. There is no gainsaying that this innovation (AI)-powered customer service engenders client loyalty [34]. The use of AI by companies in their operations better positions organizations to build long-term and profitable connections with their customers.

Consequently, integrating AI is a sure way that allows for more user-friendly self-service choices. Interactive instruments such as AI-powered information systems help clients navigate and resolve queries and/or customer concerns autonomously. A tailored solution can be obtained by customers using analytics and AI-driven predictive recommendations, thereby minimizing the possibility of a desire for direct support channels as customer service or queries are not limited by time zones or work hours, thanks to automated systems and AI chatbots. The automated systems and AI chatbots' seamless accessibility boost customer happiness, establishing a reputation for dependability and responsiveness by addressing urgent concerns quickly.

One of the main advantages of AI from the discussion is its ability to eliminate errors and improve accuracy at each stage of decision-making based on previously obtained knowledge. There is a high possibility of mistakes at a zero level when AI technology is properly programmed.

Enhancing User Experience with AI in Customer Service

Deepening empathy toward clients via AI in customer service is an arduous endeavour that leverages disparate technologies and methodologies to improve consumer interaction, satisfaction, and loyalty. Into this revolution, AI-driven solutions such as chatbots and virtual assistants are central, creating experiences for customers, simplifying service delivery and responding to queries faster.

Personalization: With AI, this is a key feature in enhancing user experience. With data values fed into AI technology, each customer's information can be processed in a way that turns each interaction into an experience tailored according to client preference and behaviour. According to Marti et al. AI can also boost customer engagement through personalization, moderation automation, and data-based insights. Such personalization makes customers feel connected with brands, eventually promoting loyalty and satisfaction. Similarly, Oyedele et al. [35] highlights how AI facilitates personalized communications through customized products or services for targeted customers, leading to a unique brand experience.

Efficient and Responsive: AI systems, particularly chatbots, can handle thousands of questions at once, offering instant responses to customer inquiries [36]. On another frontier, it showed the effectiveness of voice-driven AI systems deployed in call centres in significantly reducing complaints from knowledgeable customers, revealing the power of AI to empower service delivery. In addition, AI-driven chatbots can generate quicker and more precise answers to customer questions [37]. This ability not only enhances customer satisfaction but also enables human agents to address more intricate matters, thus enhancing operational efficiency in general.

User Acceptance and Adaptation: These are the crux of the optimal application of AI in customer support. Oyenuga et al. [38] theorize that customers would be more willing to accept chatbots if they find them delivering service at comparable levels to human employees. After the customer grows accustomed to technology, the advantage of AI on user experience is fully realized. This is of great significance for family businesses where chatbots supplement employee service, thereby improving perceived usefulness and promoting increased functional acceptance. Yet, though AI can make a difference in user experience, it must be noted that it has limitations. Folorunsho et al. [11] believe that AI-based customer service is missing the emotional intelligence needed to foster good customer relationships, especially in dealing with intricate, emotion-based needs. This underscores the need to strike a balance between AI functionality and human interaction to respond to the emotional side of customer service appropriately.

Continuous Improvement: Feedback loops are another critical approach to improving user experience with AI. Organizations can ensure continued learning and adaptation by incorporating customer interaction feedback into AI systems. Oyenuga et al. [39] support a collaborative process that incorporates AI capabilities with human knowledge, such that AI solutions adapt to changing customer needs and preferences.

Challenges and Limitations of AI-Powered Customer Service

While there are tremendous advantages to using AI-based technologies, including higher efficiency and scalability, there are also innate constraints that may impact the overall experience of the customer.

1. **Emotional and Relational Limitations:** One of the key difficulties posed by AI customer service is its inherent deficiency in handling the emotional and relational aspects of service conversations [11]. While AI can facilitate the utmost efficiency of operations, it generally lacks the emotional intelligence required to deal with customers. Services that depend on human touch, including those with complex emotional demands, will be the most difficult for AI to replicate. This is further confirmed by Najmi et al. [31], who argue that AI could assist in transactional services but that relational services are better for human encounters as they promote higher customer lifetime value.
2. **User Acceptance and Trust Issues:** Acceptance of AI technologies by the user can also be a crucial challenge. Scepticism and resistance against AI may prevent an effective service interaction [17]. The scepticism likely stems from negative experiences in the past, or simply a lack of understanding of AI tech altogether, which results in frustration and disappointment.
3. **Data Privacy and Security Issues:** AI-enabled customer support brings important data protection and privacy concerns. Most of the AI systems involve working with a significant amount of individual and sensitive data, so good protection of the data must be in place as well. Oyenuga and Fredrick [40] note that the extensive use of AI requires scrutiny of data privacy laws and regulations, as well as transparency policies of algorithms. Failure to address these issues could cause a lack of trust on behalf of the customer and legal implications. AI models can focus on different types of data, so not all application programming interfaces (APIs) will engage with you the same way at all.
4. **Quality of AI Interactions:** The average quality of interactions powered by AI wisdom can vary widely. AI is indeed capable of providing instant responses, but it lacks in terms of contextual understanding of nuances in situations, and the customer experience may leave a lot to be desired. Qiu et al. [41] point out that while AI can learn from customer satisfaction metrics to improve the service process, it may fail to understand the circumstances or mood of the customer, thus, responses can be suboptimal. This limitation can become especially problematic in industries where the satisfaction of customers is strongly related to the quality of a service interaction.
5. **Technology Dependence Problems:** Organizations which are overreliant on AI-based customer service may face technology dependence

problems. The timely and effective service delivery gets severely impaired when such AI systems break down or face technical glitches.

Also, the necessity of repeated updating and servicing of AI systems to keep them functioning optimally further adds to operating costs. In line with Rizomyliotis et al. [42], although AI can lead to increased individualization and customer satisfaction, it can potentially reduce comparability and transparency and affect customers' perceptions of fairness.

6. **Balancing AI and Human Interaction:** Finding the right balance between AI and human interaction, Sands et al. [43] advocates that customer interaction does not need to be completely automated by AI; it should act to help the human service employee, specifically within emotionally intensive and social contexts. A hybrid approach can increase the quality of service by combining the strengths of humans and AI.
7. **Ethical Considerations:** Understanding the ethical consequences of utilizing AI in customer care is also not to be dismissed. With the rising adoption of AI systems, there are issues related to bias, fairness, and transparency in decision-making. If customers feel that their needs or preferences are being captured by AI systems, they can become alienated or excluded. Shankar et al. [44] accentuates the ethical implications which need to be understood to create trust and acceptance among customers.

Strategies for Businesses to Successfully Implement AI in Customer Service

Here are some strategies that can help businesses make use of artificial intelligence to develop their customer service solutions.

Define Clear Objectives: Before implementing AI solutions, businesses must set up specific objectives and assign specific application scenarios that further their customer service mission. Yes, these could be knowing where AI will create maximum value, for instance, auto-response to basic queries, empowering greater individualized consumer interaction or just putting in the nuts and bolts to make the solutions deliver better. For example, AI-powered chatbots will create more personalized shopping experiences, but they require extensive implementation and planning [45].

Build Strong AI Infrastructure: A strong technological infrastructure is essential for the application of AI in customer support. This requires investing in the hardware, software, and data management systems that would enable AI applications. According to Singh and Singh [46], an AI-based service model can contribute greatly to optimizing operational effectiveness, but it needs a good infrastructure to operate well. Organizations need to consider cloud-based solutions and enable them to scale and evolve so that they can adapt to changing customer requirements and possible technological evolution.

Prioritize Data Quality and Management: The quality and governance of the data underpinning AI systems is paramount to their performance. Business owners require access to the correct, accurate, and real-time customer data to train their business-AI models. Folorunsho et al. [12] underlined the importance of transparent policies regarding AI services, which necessarily require a good understanding of the data that is being used. Effectively enforcing strong data governance practices ensures that organizations maintain data integrity and, consequently, the performance of their AI systems. AI tech should be used for the personalization experience of the customers.

Focus on Customer Experience and Personalization: Businesses can track customer data and behaviour, and can customize the interaction or offering based on individual preferences. According to Moriuchi [31], while AI can be used to employ transactional services, human participation must be retained for relational services. Businesses must strive to create seamless experiences that combine AI efficiency with a human touch so customers feel valued and understood.

Train Employees and Foster Collaboration: To optimize AI adoption, human workers must integrate with technology [47]. Well, companies should also invest in training programs that help employees get equipped with AI systems. As a result, this means clearly understanding the insights provided by artificial intelligence and employing them to enhance customer communication. As highlighted by Sreerangapuri [48], where the emotional and relational aspects of service matter greatly, workers must be prepared to interact with consumers in ways that are impossible for AI to do.

Track Performance and Collect Feedback: Companies should approach with key performance indicators (KPIS) on how well their AI systems are working above their expectations, and gather customer feedback on what they have encountered. According to Tula et al. [49],

AI systems can shield back-office processes from disruption, but organizations must remain vigilant to prevent service quality from deteriorating.

Solving Ethical Questions and Trust Building: As customer service gets identified with AI-based technologies, organizations are facing ethical questions around data security, privacy, and transparency. Consumer trust is being lost with the mainstream implementation of AI like Cambridge Analytica, and the "weaponization" of data, and companies have to build a foundation of trust to deliver long-term AI. An organization should be transparent about its use of AI to earn customer trust [50, 51].

Future Trends in AI-Powered Customer Service

As technology advances to match consumer expectations, the future of AI-powered customer service is going to be game-changing. Trends such as the rise of voice commerce, the rise of hyper-personalization with the support of AI and big data, the development of self-learning systems that allow for continuous improvement, and the adoption of AR (augmented reality) and VR (virtual reality) in customer service. All of these trends also offer challenges and opportunities for organizations that want to raise the bar on customer experience. With the increasing popularity of smart speakers and voice-enabled technologies, voice commerce has the potential to create seamless shopping experiences for consumers. According to Tulcanaza-Prieto et al. [52], voice assistants could help strengthen customer engagement and loyalty by giving them instant access to product information and enabling them to perform transactions using their voice. For older consumers, voice as a way to shop is an exception in that it can be an easier mental and physical way to interact with e-commerce compared to conventional interfaces [53]. Voice assistants foster emotional attachments that affect the purchasing decisions users make. Perceptions of voice assistants as family-like entities among users can improve their willingness to perform voice commerce. With the evolution of voice technology organizations need to refine their strategy of voice commerce to capitalize on this trend [54].

One other important trend in AI-powered customer service is hyper-personalization, which is made possible by the enormous amounts of data generated per day by consumer interactions. With big data analytics, enterprises can customize experiences to serve customers at an individual level. Online personalization takes this step further, implementing AI algorithms to analyze a consumer's behaviour, preference, and buying history on a real-time basis [55]. For the insights that they provide, personalized recommendations and targeted marketing messages can greatly improve customer satisfaction and loyalty. For instance, AI-driven systems can analyze user data and predict future needs and preferences of users, and as a result, businesses would be able to preemptively reach out to their customers with offers that align with their demands [55, 56]. Advanced personalization undoubtedly provides a better customer experience [57, 58] and eventually helps in increasing conversion rates and customer retention [59], and driving revenue growth as a result [60]. Furthermore, through artificial intelligence self-learning techniques, which are the leading customer service technology, we can observe an increasing perception. These systems utilize machine learning algorithms, which allow them to continuously learn from performance data and customer feedback and adapt over time for continuous improvement [61]. Self-learning systems can help organizations improve their service quality and operational efficiency by determining the key areas in which improvement is needed and adjusting processes accordingly. These techniques are driven by self-learning systems that can observe customer interactions and identify the most commonplace pain points, enabling the business to proactively respond and improve service with each interaction. This is the best evidence of how institutions can cater demands of buyers and interests directly for keeping customer service strategies alive [62].

Moreover, customer support methodologies incorporating AR and VR innovations provide immersive events that result in enhanced user interaction [7]. Such technologies allow customers to visualize products in their actual residence or services in a hyper-realistic setting, thereby connecting online with offline experience [63].

Contrary to that, AR applications allow customers to try clothes rather than going out to a shop or settle furniture in their home thus helping them to make more convenient choices in buying [64]. Also, AR and VR give great direction to users by providing interactive tutorials or arranging fixes. Organizations rely on immersive training experiences to empower customers to self-service solutions, thus alleviating the burden on customer service agents [64].

Impacts of AI-Powered Customer Service

The overall customer experience is greatly enhanced through the use of AI-powered customer service, as customers receive personalized and prompt attention. According to Davenport et al. [9], customers have a positive effect concerning benefits of convenience, personalization, and trust; positive attitude toward AI-enabled services. Such customizations glorify brand loyalty and touch points as customers feel noticed and engaged. Furthermore, AI-based chatbots and virtual assistants work constantly for 24 hours, allowing customers to query support whenever they want [65]. This immediacy not only makes for a more rewarding customer experience, but it also eliminates frustration with the prospect of having to wait long periods for a human agent. Oyedele and Opeyemi [36] explain that, while a voice-enabled AI system can fulfil customer requests, the extent to which that is possible depends on the nature of the requests and how familiar the user is with the AI. Using AI the smart way in customer support can give businesses a huge boost in operational efficiencies.

Additionally, businesses can devote human resources to more complex and nuanced interactions with customers through the use of this technology for automating mundane tasks and queries. With this change, organizations can optimize their human capital, enhancing the quality of the services they deliver. According to Zada [66], AI amenities enhance service quality because they free up employees for more important tasks, thereby boosting customer engagement. AI is also enabling the systems to learn from experience and continue to improve their skills. Following the self-adapting characteristic of artificial intelligence, organizations can delight in changing customer needs and wants. Robotic technologies, which chatbots are part of, can reduce customer service costs by nearly 30% [67]. The need for massive human capital is reduced since these systems can respond to and oversee a large number of questions at once, leading to reduced operational expenses. This monetary productivity is especially crucial to organizations that want to stay competitive in a dynamically changing market.

The more intelligent the AI systems get, the more they can strengthen relationships with customers by delivering personalized answers and engaging proactively. Zahra et al. [67] propose that AI service robots can stimulate customer interactions, and enhance engagement by providing reliable and accurate service according to their needs. Such evolution in customer relations can translate into enhanced customer loyalty and a long-term brand ambassador. That said, it is important to remember that while AI can augment customer engagement, it cannot entirely replace the human element. Moriuchi [31] avow that humans are the superior agents for relational services that need emotional intelligence and empathy. So, businesses should balance the efficiency offered by AI with the human touch that is needed to keep customer interactions high-quality.

Conclusion

This application of artificial intelligence in customer service is a game changer in the relationship of a customer with a business. AI-based solutions can add value to organizations, enabling a better customer experience, improving operational efficiency, reducing costs and tracking customer relationships. AI is highly promising, but organizations will need to tread carefully, ensuring that customer engagement is still human-oriented, while addressing ethical issues linked to data security and privacy. AI is an advancing technology, and the rapid growth of AI technologies will continue to bring innovations in customer service in the coming years. Only organizations that will adjust to the changes and adjust their directions as per the changes will survive in such complicated paths.

References

1. Abufawr, M., Alawami, A., Attar, M., Obaid, A., Alharbi, M., Alawami, M., … Alfahke, H. (2024). Evaluation of the role of artificial intelligence on customer satisfaction as a competitive advantage in the hospitality industry in the United States. *Technology and Investment*, 15(03), 135–144. https://doi.org/10.4236/ti.2024.153008
2. Adam, M., Wessel, M., & Benlian, A. (2020). AI-based chatbots in customer service and their effects on user compliance. *Electronic Markets*, 31(2), 427–445. https://doi.org/10.1007/s12525-020-00414-7
3. Alam, M. N., Turi, J. A., Bhuiyan, A. B., Kharusi, S. A., Oyenuga, M., Zulkifli, N., & Iqbal, J. (2024). Factors influencing intention for reusing virtual reality (VR) at theme parks: The mediating role of visitors' satisfaction. *Cogent Social Sciences*, 10(1). https://doi.org/10.1080/23311886.2023.2298898
4. Al-Jedibi, W. (2022). The strategic plan of the information technology deanship—King Abdulaziz University—Saudi Arabia. *International Journal of Applied Information Management*, 2, 84–94.
5. Al-Shoteri, A. (2022). The role of methods and applications of artificial intelligence tools in the field of medicine to diagnose and discover various diseases. *Journal of Applied Data Science*, 3(1), 1–14
6. Bhuiyan, M. (2024). The role of AI-enhanced personalization in customer experiences. *Journal of Computer Science and Technology Studies*, 6(1), 162–169. https://doi.org/10.32996/jcsts.2024.6.1.17

7. Brzozowska, M., Kolasińska-Morawska, K., Sułkowski, Ł., & Morawski, P. (2023). Artificial-intelligence-powered customer service management in the logistics industry. *Entrepreneurial Business and Economics Review*, 11(4), 109–121. https://doi.org/10.15678/eber.2023.110407
8. Castillo, D., Canhoto, A., & Said, E. (2020). The dark side of AI-powered service interactions: Exploring the process of co-destruction from the customer perspective. *Service Industries Journal*, 41(13–14), 900–925. https://doi.org/10.1080/02642069.2020.1787993
9. Davenport, T. H., Guha, A., Grewal, D., & Bressgott, T. (2020). How artificial intelligence will change the future of marketing. *Journal of the Academy of Marketing Science*, 48(1), 24–42.
10. El-Shihy, D., Abdelraouf, M., Hegazy, M., & Hassan, N. (2024). The influence of AI chatbots in fintech services on customer loyalty within the banking industry. *Future of Business Administration*, 3(1), 16–28. https://doi.org/10.33422/fba.v3i1.644
11. Fan, H., Han, B., Gao, W., & Li, W. (2022). How AI chatbots have reshaped the frontline interface in China: Examining the role of sales–service ambidexterity and the personalization–privacy paradox. *International Journal of Emerging Markets*, 17(4), 967–986. https://doi.org/10.1108/ijoem-04-2021-0532
12. Folorunsho, S., Adenekan, O., Ezeigweneme, C., Somadina, I., & Okeleke, P. (2024). Leveraging technical support experience to implement effective AI solutions and future service improvements. *International Journal of Applied Research in Social Sciences*, 6(8), 1758–1783. https://doi.org/10.51594/ijarss.v6i8.1425
13. Følstad, A., & Brandtzæg, P. (2020). Users' experiences with chatbots: Findings from a questionnaire study. *Quality and User Experience*, 5(1). https://doi.org/10.1007/s41233-020-00033-2
14. George, A. S., & George, A. H. (2023). A review of ChatGPT AI's impact on several business sectors. *Partners Universal International Innovation Journal*, 1(1), 9–23. https://doi.org/10.5281/zenodo.7644359
15. Ghosh, S., Ness, S., & Salunkhe, S. (2024). The role of AI-enabled chatbots in omnichannel customer service. *Journal of Engineering Research and Reports*, 26(6), 327–345. https://doi.org/10.9734/jerr/2024/v26i61184
16. Han, Y., Chen, J., Dou, M., Wang, J., & Feng, K. (2023). The impact of artificial intelligence on the financial services industry. *Academic Journal of Management and Social Sciences*, 2(3), 83–85. https://doi.org/10.54097/ajmss.v2i3.8741
17. Henkel, A., Bromuri, S., İren, D., & Urovi, V. (2020). Half human, half machine – augmenting service employees with AI for interpersonal emotion regulation. *Journal of Service Management*, 31(2), 247–265. https://doi.org/10.1108/josm-05-2019-0160
18. Hitoshi, H. (2021). The effectiveness of the body of knowledge process in the startup analysis of efficiency by applying startup management body of knowledge (SUBOK) guide. *International Journal of Applied Information Management*, 1(2), 28–49.

19. Ho, S., & Chow, M. (2023). The role of artificial intelligence in consumers' brand preference for retail banks in Hong Kong. *Journal of Financial Services Marketing*, 29(2), 292–305. https://doi.org/10.1057/s41264-022-00207-3
20. Huang, M., & Rust, R. (2018). Artificial intelligence in service. *Journal of Service Research*, 21(2), 155–172. https://doi.org/10.1177/1094670517752459
21. Huang, Y., & Gürsoy, D. (2024). Customers' online service encounter satisfaction with chatbots: Interaction effects of language style and decision-making journey stage. *International Journal of Contemporary Hospitality Management*, 36(12), 4074–4091. https://doi.org/10.1108/ijchm-11-2023-1800
22. Hussain, A., Khan, M., Manoj, B., & Shalini, K. (2024). *A study on the impact of artificial intelligence in financial services of private banks in Bangalore*, 91–98. https://doi.org/10.48001/978-81-966500-8-7-9
23. Jiang, K., Qin, M., & Li, S. (2022). Chatbots in retail: How do they affect the continued use and purchase intentions of Chinese consumers? *Journal of Consumer Behaviour*, 21(4), 756–772. https://doi.org/10.1002/cb.2034
24. Kanapathipillai, K., Nian, O., Yi, C., & Ping, Y. (2024). From service to superiority: Unveiling the impact of artificial intelligence on customer experience in giant hypermarkets, Klang Valley, Malaysia. *European Journal of Management and Marketing Studies*, 9(2). https://doi.org/10.46827/ejmms.v9i2.1844
25. Karamthulla, M., Narkarunai, J., Malaiyappan, A., & Prakash, S. (2023). Ai-powered self-healing systems for fault-tolerant platform engineering: Case studies and challenges. *Journal of Knowledge Learning and Science Technology*, 2(2), 327–338. https://doi.org/10.60087/jklst.vol2.n2.p338
26. Khana, A., Hamid, A., Saad, N., & Arif, A. (2023). Effectiveness of artificial intelligence in building customer loyalty: Investigating the mediating role of chatbot in the tourism sector of Pakistan. *International Journal of Academic Research in Business and Social Sciences*, 13(9). https://doi.org/10.6007/ijarbss/v13-i9/18422
27. Marti, C., Liu, H., Kour, G., Bilgihan, A., & Xu, Y. (2024). Leveraging artificial intelligence in firm-generated online customer communities: A framework and future research agenda. *Journal of Service Management*, 35(3), 438–458. https://doi.org/10.1108/josm-10-2023-0443
28. Mehrotra, A. (2019). *Artificial intelligence in financial services-need to blend automation with human touch* [Paper presentation]. 2019 International Conference on Automation, Computational and Technology Management (ICACTM), pp. 342–347. IEEE. https://doi.org/10.1109/ICACTM.2019.8776741
29. Mohanty, A., Mohanty, J., Lingam, N., & Acharya, A. (2023). Artificial intelligence transforming customer service management: Embracing the future. *The Oriental Studies*, 12(2), 35–47.
30. Meng, L., Chen, J., Yang, M., & Wang, Y. (2024). Effects of customer inoculation on artificial intelligence service failure. *International Journal of Contemporary Hospitality Management*, 37(2), 444–461. https://doi.org/10.1108/ijchm-01-2024-0140

31. Moriuchi, E. (2019). Okay, Google!: An empirical study on voice assistants on consumer engagement and loyalty. *Psychology and Marketing*, 36(5), 489–501. https://doi.org/10.1002/mar.21192
32. Najmi, A., Kanapathy, K., & Aziz, A. A. (2021). Understanding consumer participation in managing ICT waste: Findings from two-staged structural equation modeling–artificial neural network approach. *Environmental Science and Pollution Research*, 28, 14782–14796.
33. Nguyen, M., & Malik, A. (2021). Impact of knowledge sharing on employees' service quality: The moderating role of artificial intelligence. *International Marketing Review*, 39(3), 482–508. https://doi.org/10.1108/imr-02-2021-0078
34. Ntumba, C., Aguayo, S., & Maina, K. (2023). Revolutionizing retail: A mini-review of e-commerce evolution. *Journal of Digital Marketing and Communication*, 3(2), 100–110. https://doi.org/10.53623/jdmc.v3i2.365
35. Oyedele, M.O., Sunday, A.O., & Abuh, A.I. (2024). Fostering technological-enhanced training and development for business survival and performance in the new normal. *Journal of Propulsion Technology*, 45(3), 1858–1869. https://propulsiontechjournal.com/index.php/journal/article/view/7479
36. Oyedele, O. M., & Opeyemi, L. (2024). Using marketing mix strategies to influence customer patronage in an indigenous restaurant. *Rowter Journal*, 3(1), 35–50.
37. Oyenuga, M.O., Akhaine, M. E., & Omale, S.A. (2023). We need it: How digital marketing tactics influence the purchasing behaviour of Nigerian Millenials. *Konfrontasi Journal: Culture, Economy and Social Changes*, 10(2), 91–105 https://doi.org/10.33258/konfrontasi2.v10i2.275
38. Oyenuga, M. O., Jelili, B., & Bulugbe, K. (2020). E-banking and customer satisfaction in the banking industry. *Nigeria Journal of Business Administration*, 18(1), 112–132. https://api-ir.unilag.edu.ng/server/api/core/bitstreams/53eb4271-c536-4c99-8910-64622549ba98/content#page=120
39. Oyenuga, M. O., Andah, R. A., & Nduji, R. (2023). Employee participation and organisational performance in an emerging economy. *Konfrontasi Journal: Culture, Economy and Social Changes*, 10(2), 91–105 https://doi.org/10.33258/konfrontasi2.v10i3.281
40. Oyenuga, M. O., & Fredrick, E. (2023). Managing product features and customer satisfaction via product development, product branding, and product packaging: Evidence from CWAY Table Water. *Marketing and Branding Research*, 10(1), 28–44. https://doi.org/10.32038/mbr.2023.10.01.03
41. Qiu, H., Li, M., Bai, B., Wang, N., & Li, Y. (2022). The impact of AI-enabled service attributes on service hospitableness: The role of employee physical and psychological workload. *International Journal of Contemporary Hospitality Management*, 34(4), 1374–1398. https://doi.org/10.1108/ijchm-08-2021-0960
42. Rizomyliotis, I., Kastanakis, M., Giovanis, A., Konstantoulaki, K., & Kostopoulos, I. (2022). "How may I help you today?" The use of AI chatbots in small family businesses and the moderating role of customer affective commitment. *Journal of Business Research*, 153, 329–340. https://doi.org/10.1016/j.jbusres.2022.08.035

43. Sands, S., Ferraro, C., Campbell, C., & Tsao, H. (2020). Managing the human–chatbot divide: How service scripts influence service experience. *Journal of Service Management*, 32(2), 246–264. https://doi.org/10.1108/josm-06-2019-0203
44. Shankar, G., Suresh, M., & Lakshmi, T. (2024). Application of chatbots and virtual assistants in ticket booking system. *International Research Journal of Advanced Engineering and Management,* 2(05), 1605–1608. https://doi.org/10.47392/irjaem.2024.0221
45. Shah, T., Kautish, P., & Mehmood, K. (2023). Influence of robots service quality on customers' acceptance in restaurants. *Asia Pacific Journal of Marketing and Logistics*, 35(12), 3117–3137. https://doi.org/10.1108/apjml-09-2022-0780
46. Singh, P., & Singh, V. (2024). The power of AI: Enhancing customer loyalty through satisfaction and efficiency. *Cogent Business & Management*, 11(1). https://doi.org/10.1080/23311975.2024.2326107
47. Spring, M., Faulconbridge, J., & Sarwar, A. (2022). How information technology automates and augments processes: Insights from artificial-intelligence-based systems in professional service operations. *Journal of Operations Management*, 68(6–7), 592–618. https://doi.org/10.1002/joom.1215
48. Sreerangapuri, A. (2024). Ai-driven service transformation: Revolutionizing operational excellence. *International Journal of Scientific Research in Computer Science Engineering and Information Technology*, 10(6), 132–140. https://doi.org/10.32628/cseit24106154
49. Tula, S., Kess-Momoh, A., Omotoye, G., Bello, B., & Daraojimba, A. (2024). Ai-enabled customer experience enhancement in business. *Computer Science & It Research Journal*, 5(2), 365–389. https://doi.org/10.51594/csitrj.v5i2.789
50. Tulcanaza-Prieto, A., Cortez-Ordoñez, A., & Lee, C. (2023). Influence of customer perception factors on AI-enabled customer experience in the Ecuadorian banking environment. *Sustainability*, 15(16), 12441. https://doi.org/10.3390/su151612441
51. Tsai, W., Liu, Y., & Chuan, C. (2021). How chatbots' social presence communication enhances consumer engagement: The mediating role of parasocial interaction and dialogue. *Journal of Research in Interactive Marketing*, 15(3), 460–482. https://doi.org/10.1108/jrim-12-2019-0200
52. Tulcanaza-Prieto, A., Cortez-Ordoñez, A., & Lee, C. (2023). Influence of customer perception factors on AI-enabled customer experience in the Ecuadorian banking environment. *Sustainability*, 15(16), 12441. https://doi.org/10.3390/su151612441
53. Trawnih, A., Al-Masaeed, S., Alsoud, M., & Alkufahy, A. (2022). Understanding artificial intelligence experience: A customer perspective. *International Journal of Data and Network Science*, 6(4), 1471–1484. https://doi.org/10.5267/j.ijdns.2022.5.004
54. Uzoka, N. A., Cadet, N. E., & Pascal, N. (2024). Leveraging AI-powered chatbots to enhance customer service efficiency and future opportunities in automated support. *Computer Science & IT Research Journal*, 5(10), 2485–2510. https://doi.org/10.51594/csitrj.v5i10.1676

55. Voicu, M., Sîrghi, N., & Toth, D. (2023). Consumers' experience and satisfaction using augmented reality apps in e-shopping: New empirical evidence. *Applied Sciences*, 13(17), 9596. https://doi.org/10.3390/app13179596
56. Wang, L., Huang, N., Hong, Y., Liu, L., Guo, X., & Chen, G. (2023). Voice-based AI in call centre customer service: A natural field experiment. *Production and Operations Management*, 32(4), 1002–1018. https://doi.org/10.1111/poms.13953
57. Whang, C., & Im, H. (2020). "I like your suggestion!" the role of human likeness and parasocial relationship on the website versus voice shopper's perception of recommendations. *Psychology and Marketing*, 38(4), 581–595. https://doi.org/10.1002/mar.21437
58. Wang, L., Huang, N., Hong, Y., Liu, L., Guo, X., & Chen, G. (2023). Voice-based AI in call centre customer service: A natural field experiment. *Production and Operations Management*, 32(4), 1002–1018. https://doi.org/10.1111/poms.13953
59. Weith, H., & Matt, C. (2022). When do customers perceive artificial intelligence as fair? An assessment of AI-based b2c e-commerce services. https://doi.org/10.24251/hicss.2022.529
60. Wirtz, J., So, K. K. F., Mody, M. A., Liu, S. Q., & Chun, H. H. (2019). Platforms in the peer-to-peer sharing economy. *Journal of Service Management*, 30(4), 452–483.
61. Xiao, L., & Kumar, V. (2019). Robotics for customer service: A useful complement or an ultimate substitute? *Journal of Service Research*, 24(1), 9–29. https://doi.org/10.1177/1094670519878881
62. Yang, X. (2023). The effects of AI service quality and AI function-customer ability fit on customer's overall co-creation experience. *Industrial Management & Data Systems*, 123(6), 1717–1735. https://doi.org/10.1108/imds-08-2022-0500
63. Yeo, S. F., Tan, C. L., Kumar, A., Tan, K. H., & Wong, J. K. (2022). Investigating the impact of AI-powered technologies on Instagrammers' purchase decisions in digitalization era–A study of the fashion and apparel industry. *Technological Forecasting and Social Change*, 177, 121551. https://doi.org/10.1016/j.techfore.2022.121551
64. Yu, X., Xu, Z., Song, Y., & Liu, X. (2022). The cuter, the better. the impact of cuteness on intention to purchase AI voice assistants: A moderated serial-mediation model. *Frontiers in Psychology*, 13. https://doi.org/10.3389/fpsyg.2022.1036848
65. Zhou, T. Y., Liu, Y., & Kumar, M. (2024). *AI-powered Chatbots for improving interactive user experience: State-of-the-art.* CIE51 Proceedings, pp. 1–13.
66. Zada, I. (2022). The contributions of customer knowledge and artificial intelligence to customer satisfaction. *International Review of Management and Marketing*, 12(5), 1–4. https://doi.org/10.32479/irmm.13314
67. Zahra, A., Jonas, D., Erliyani, I., Rosdiana, R., & Yusuf, N. (2023). Assessing customer satisfaction in AI-powered services: An empirical study with smart pls. *International Transactions on Artificial Intelligence (Italic)*, 2(1), 81–89. https://doi.org/10.33050/italic.v2i1.432

Chapter 14

Case Study on AI-Powered Healthcare: Improving Patient Outcomes with Predictive Analytics

Gulam Mahabub Subhani and Ch. Sumalakshmi

Introduction

The integration of AI-driven analytics in healthcare has the potential to significantly enhance decision-making and operational efficiency. By utilizing artificial intelligence and sophisticated data analysis techniques, healthcare organizations can extract meaningful insights from vast and complex datasets. These insights contribute to improved patient care, cost reduction, and streamlined operations. Understanding the various applications of AI-powered analytics is essential in recognizing its role in advancing decision-making processes within healthcare systems. One of the most impactful applications of AI-driven analytics is predictive analytics. By analyzing historical patient data, including electronic health records (EHRs), medical imaging, genetic information, and wearable device metrics, AI algorithms can identify patterns and generate predictions regarding health outcomes. This predictive capability enables healthcare professionals to detect and

DOI: 10.4324/9781003606956-14

address diseases at an early stage, optimize treatment strategies, and allocate resources more efficiently.

AI also plays a crucial role in clinical decision support by providing real-time recommendations at the point of care. These intelligent systems analyze patient information and medical literature to assist healthcare providers in diagnosing complex conditions, recommending appropriate treatments, and identifying potential drug interactions or adverse effects. By integrating AI-powered decision support systems, clinicians can make well-informed decisions promptly, improving overall patient safety and quality of care [1, 2]. In addition to clinical decision-making, AI-driven analytics significantly enhances resource management within healthcare facilities. By assessing patient flow, bed occupancy rates, and staff schedules, AI algorithms can forecast patient demand and optimize resource distribution. This leads to reduced wait times, improved patient satisfaction, and enhanced operational efficiency. Addressing healthcare fraud is another critical area where AI analytics demonstrates its value. AI algorithms can analyze vast amounts of healthcare claims data to detect patterns indicative of fraudulent activities. This capability helps prevent fraud, reducing financial losses for insurance providers and patients while strengthening trust in the healthcare system.

AI is also transforming drug discovery and development by rapidly analyzing scientific literature, clinical trial data, and molecular structures. This accelerates the identification of potential drug candidates, predicts their efficacy, and optimizes clinical trial design, ultimately reducing the time and cost required to develop new treatments. This advancement enhances patient access to innovative therapies and revolutionizes pharmaceutical research [3, 4]. Moreover, AI contributes to pharmacovigilance efforts by analyzing adverse event reports and other relevant data sources to detect potential safety signals. By prioritizing these signals based on significance, regulatory agencies and healthcare organizations can efficiently allocate resources for further investigation and decision-making. In the field of maternal and fetal health, AI-powered analytics plays a crucial role in fetal monitoring and diagnostics [5]. By analyzing fetal heart rate patterns and uterine activity, AI algorithms can detect potential complications and provide real-time alerts to healthcare providers. This enables timely interventions that safeguard the health of both expectant mothers and their babies [6].

The implementation of AI-powered analytics in healthcare holds immense promise for improving decision-making and operational efficiency across multiple domains. By leveraging artificial intelligence and advanced data analysis, healthcare organizations can unlock valuable insights that lead to better patient outcomes, cost savings, and optimized processes. The transformative impact of AI is evident in predictive analytics, clinical decision support, resource management, fraud detection, drug discovery, pharmacovigilance, and fetal monitoring. Embracing AI-driven technologies will shape the future of healthcare, benefiting patients, medical professionals, and society as a whole [7].

The Role of Predictive Analytics in Healthcare

Predictive analytics is transforming the healthcare landscape by harnessing vast amounts of patient data, including electronic health records (EHRs), medical imaging, genetic profiles, and data from wearable health devices. Through sophisticated algorithms and machine learning techniques, healthcare providers can analyze these datasets to detect patterns that help anticipate patient health trends. This approach enhances medical care by facilitating early disease identification, refining treatment strategies, and streamlining resource management [8, 9].

By examining historical patient records, predictive analytics assists in the early diagnosis and prevention of diseases. It can uncover subtle indicators within EHRs and genetic information that signal potential health risks. Identifying these risks at an early stage enables healthcare professionals to take proactive measures, reducing the likelihood of disease progression and improving overall patient well-being [10]. This method helps medical teams focus on individuals with a higher risk of developing certain conditions, ensuring timely interventions and improving the efficiency of healthcare delivery.

A key advantage of predictive analytics is its role in optimizing treatment plans. By evaluating a broad spectrum of patient data, including genetic markers, medical images, and previous treatment responses, predictive models can identify trends that guide clinicians toward the most effective therapeutic options. This personalized approach enhances treatment precision, increasing the likelihood of successful outcomes while minimizing unnecessary interventions. Data-driven insights support healthcare professionals in selecting appropriate therapies, adjusting medication doses, and determining optimal treatment durations, all of which contribute to improved patient care. Another significant benefit of predictive analytics is its contribution to healthcare resource management. By analyzing past data related to patient admissions, disease prevalence, and demographic trends, predictive models can forecast future healthcare demands. This enables medical facilities to allocate resources more effectively, ensuring that patients receive prompt and adequate care [11]. Efficient resource distribution leads to shorter wait times, increased patient satisfaction, and improved operational effectiveness. By optimizing resource utilization, healthcare providers can maintain high-quality services while reducing costs.

The integration of predictive analytics into healthcare is paving the way for a more data-driven and proactive approach to medical care [12]. By generating meaningful insights from diverse data sources, this technology aids in early disease detection, enhances treatment planning, and improves resource allocation [13]. As advancements in analytics continue, predictive modeling will play an increasingly vital role in fostering a more efficient, personalized, and cost-effective healthcare system [14].

Support for Making Clinical Decisions

Clinical decision support systems (CDSS) have become an essential tool in modern healthcare, significantly enhancing how medical professionals assess and manage patient care. These systems utilize sophisticated algorithms to analyze extensive patient records and medical literature, offering timely recommendations that improve diagnostic accuracy, particularly in complex cases. By synthesizing diverse data sources, they enable healthcare providers to evaluate patient conditions more efficiently and make well-informed clinical decisions [15].

Beyond assisting in diagnostics, CDSS play a pivotal role in guiding treatment plans. By analyzing patient-specific information, medical guidelines, and research-based evidence, these systems suggest the most appropriate therapeutic options. With access to a vast repository of medical knowledge, healthcare professionals can deliver more precise and personalized treatment plans. Additionally, these systems ensure that clinical decisions remain aligned with the latest medical research and best practices, keeping practitioners up to date with evolving standards of care. A key advantage of CDSS is their ability to detect potential drug interactions and adverse effects. By cross-referencing a patient's medical history, prescribed medications, and risk factors, these systems serve as a critical safeguard against harmful reactions [16]. This proactive approach significantly enhances patient safety by reducing the likelihood of medication-related complications.

Another notable benefit is the system's capacity to evolve continuously. As new medical research, clinical guidelines, and patient outcomes become available, CDSS refine their recommendations to reflect the latest advancements. This adaptability ensures that healthcare providers have access to the most current and reliable information, strengthening their decision-making process [17]. By incorporating these intelligent tools into clinical practice, healthcare providers can minimize diagnostic errors, optimize treatment strategies, and improve overall patient safety. The ability to process vast amounts of medical data allows for better clinical outcomes, higher quality care, and more efficient healthcare delivery. As technology advances, CDSS will continue to shape the future of medical decision-making, offering new possibilities for enhanced patient care [18, 19].

Enhancing Resource Allocation in Healthcare with AI

Efficient resource distribution is essential for effective healthcare operations. By leveraging AI-powered analytics, medical institutions can better manage their resources by analyzing vast amounts of data, including patient admissions, bed availability, and staff scheduling. These predictive insights help healthcare administrators foresee patient demand and allocate resources accordingly, ensuring smoother day-to-day operations.

With accurate predictions of patient influx, hospitals and clinics can strategically assign critical resources such as beds, medical personnel, and equipment based on both current and anticipated needs. This proactive planning reduces patient wait times, enhances overall satisfaction, and streamlines healthcare services [20].

AI-driven analytics also offer valuable information on usage trends, enabling facilities to improve resource efficiency. By identifying peak demand times, administrators can optimize staff schedules, preventing workforce shortages or workflow disruptions. This data-informed approach not only boosts operational efficiency but also guarantees timely and effective patient care. Beyond improving daily operations, smart resource management helps cut down on unnecessary expenses caused by underutilized equipment or excessive staffing. AI-powered insights reveal inefficiencies, allowing healthcare leaders to make well-informed decisions that balance cost-effectiveness with high-quality service.

By directing resources to where they are needed most—such as critical care units or high-demand departments—healthcare facilities can enhance service delivery while maintaining financial sustainability. The integration of AI in resource management serves as a game-changing strategy, fostering data-driven decision-making that improves patient care and ensures long-term operational efficiency [21]. Adopting AI for resource optimization is a crucial step toward building a more responsive, effective, and financially sustainable healthcare system.

Identifying Medical Fraud

Detecting fraud in the healthcare sector is a significant challenge, requiring advanced techniques to identify and prevent deceptive activities. The integration of artificial intelligence (AI) into fraud detection processes has introduced a transformative approach. By leveraging sophisticated algorithms that analyze extensive healthcare claims data, AI systems can recognize complex patterns and detect signs of fraudulent behavior. Unlike conventional manual review methods, which may overlook subtle anomalies, AI-powered analytics can swiftly and accurately identify suspicious claims [22]. This proactive approach helps mitigate financial losses for insurers and protects patients from potential harm associated with fraudulent medical practices.

Given the vast volume of healthcare claims processed daily, traditional fraud detection methods often prove inefficient, time-consuming, and prone to human error [23]. AI-driven systems, on the other hand, possess the ability to rapidly process and examine large datasets, identifying inconsistencies and unusual patterns that may indicate fraudulent activities. This enables real-time detection and intervention, making fraud prevention more effective. The financial consequences of healthcare fraud are substantial, leading to increased costs for insurance companies and higher premiums for policyholders [24, 25]. By adopting AI-driven fraud detection, healthcare organizations can significantly reduce financial losses, preserving the economic stability of the industry while preventing unnecessary cost burdens on patients.

Beyond financial concerns, healthcare fraud also affects the quality and efficiency of medical services. Fraudulent claims divert resources away from genuine patient care, leading to delays and reduced service quality. AI-enhanced fraud detection ensures that resources are allocated effectively, allowing healthcare providers to focus on legitimate medical needs. By optimizing resource distribution, healthcare facilities can enhance operational efficiency, reduce costs linked to fraudulent activities, and improve overall patient care.

Additionally, AI-driven fraud detection serves as a deterrent to potential fraudsters. The awareness that AI continuously monitors and analyzes claims data discourages fraudulent practices before they even begin [26]. This preventive aspect strengthens the security of the healthcare system, fostering trust among stakeholders and reinforcing the industry's integrity.

The adoption of AI in fraud detection represents a groundbreaking advancement in the healthcare sector. By utilizing advanced algorithms to detect fraud more effectively than traditional methods, AI-driven analytics provide numerous benefits, including financial protection, improved patient care, better resource management, and a robust deterrence mechanism against fraudulent activities. Integrating AI into fraud prevention strategies is essential for maintaining the credibility and sustainability of the healthcare system [27].

Medicinal Discovery and Development

The process of discovering and developing new medications is intricate, requiring extensive resources, time, and financial investment. Recent advancements in artificial intelligence (AI) have introduced transformative changes to this field, allowing for the efficient analysis of vast amounts of scientific data, clinical trial results, and molecular structures. AI-driven tools equip researchers with critical insights, expediting various phases of drug development [28].

By leveraging AI, scientists can quickly identify promising drug candidates that address urgent medical challenges. AI algorithms assess scientific literature, patents, and other relevant datasets to pinpoint molecules or compounds with potential therapeutic benefits. This capability enhances the initial phase of drug discovery, providing a strong foundation for further research and testing.

AI also improves the precision of predicting a drug's effectiveness. Through machine learning models, AI can analyze complex biological, genetic, and clinical data to determine how a compound interacts with specific targets or pathways [29]. This allows researchers to prioritize the most viable candidates, streamlining development and improving efficiency.

Furthermore, AI contributes to optimizing clinical trials, a critical step in determining the safety and efficacy of new treatments. Conducting these trials is often costly and time-consuming, but AI-powered algorithms can evaluate past trial data—including patient characteristics, treatment methods, and outcomes—to identify significant trends [30, 31]. This aids in refining study protocols, selecting suitable patient groups, and setting clear objectives, ultimately reducing costs and increasing trial success rates.

Additionally, AI enhances drug design by analyzing molecular structures, including protein compositions and chemical properties. Advanced computational models predict interactions between drug candidates and biological targets, enabling the development of compounds with stronger binding properties and fewer side effects [32]. This accelerates the transition from preclinical research to human testing, improving the overall drug development timeline. The integration of AI in drug discovery and development has significantly improved efficiency, reducing both time and financial constraints. By identifying potential treatments, enhancing efficacy predictions, and refining clinical trial processes, AI is reshaping pharmaceutical research. These advancements hold great potential for addressing unmet medical needs, fostering innovative therapies, and ultimately improving patient care on a global scale [33].

Improving Signal Detection and Goal Setting in Pharmacovigilance

Signal detection and prioritization in pharmacovigilance are essential processes for ensuring the safety and effectiveness of pharmaceutical products and healthcare interventions. The incorporation of artificial intelligence (AI)-driven analytics has revolutionized these areas by leveraging sophisticated algorithms and data analysis techniques [34]. AI can efficiently examine data from the FDA Adverse Event Reporting System (FAERS) and other relevant sources to identify patterns and potential safety concerns that require further investigation. This approach extends beyond analyzing individual adverse event reports by uncovering underlying trends and relationships that contribute to a more comprehensive understanding of drug safety.

By employing advanced analytics within AI-powered frameworks, healthcare organizations and regulatory agencies can enhance their ability to prioritize safety signals effectively. This prioritization process considers multiple factors, including the severity of reported adverse reactions, frequency of occurrences, and possible connections to specific drug categories or patient demographics. Such a structured methodology ensures that critical signals with the most significant potential impact on public health receive appropriate attention and thorough evaluation. Efficient allocation of resources is thus optimized, focusing on cases that require urgent intervention [35, 36].

Utilizing AI in pharmacovigilance offers substantial advantages. Automated AI algorithms can process and analyze vast amounts of data more efficiently than manual methods, significantly increasing scalability. This expanded capability enables a broader and more detailed assessment of potential safety concerns, improving the chances of detecting emerging risks early. Furthermore, AI-powered analytics can identify subtle trends and correlations that traditional methods might overlook. Through machine learning techniques, AI can discern complex interactions between drug exposure and adverse effects, providing deeper insights into potential risks and previously unrecognized interactions [37, 38]. These insights contribute to informed decisions regarding drug labeling, post-market surveillance, and risk management strategies [39].

Another major benefit of integrating AI-driven analytics in signal detection and prioritization is the enhancement of real-time monitoring and response. AI algorithms can rapidly process and evaluate new data, enabling swift detection of emerging safety concerns. This capability allows regulatory bodies and healthcare providers to take proactive measures in addressing potential risks, minimizing harm to patients, and maintaining public trust in the healthcare system [40], [41].

Moreover, the use of AI in pharmacovigilance encourages collaboration and knowledge-sharing among healthcare stakeholders. Standardized approaches to signal evaluation ensure consistency in decision-making across regulatory agencies and pharmaceutical organizations. Such harmonization enhances efficiency, facilitates best practice exchange, and promotes shared learning, ultimately improving global patient safety outcomes [42]. In conclusion, integrating AI-driven analytics in signal detection and pharmacovigilance prioritization represents a significant advancement in drug safety monitoring. AI technologies enable efficient analysis of FAERS data and other sources, aiding in the identification and prioritization of potential safety concerns. This approach enhances resource allocation efficiency, supports timely recognition of emerging risks, detects intricate patterns and associations, and fosters collaboration within the healthcare industry. Adopting AI-driven solutions in pharmacovigilance offers immense potential for strengthening patient safety and enhancing public health protection [43, 44].

Monitoring and Evaluation of the Fetus

Ensuring the health and safety of both expectant mothers and their unborn babies is a top priority in prenatal care. Advances in analytical technology have significantly enhanced the accuracy and effectiveness of fetal monitoring and diagnostics. By closely tracking essential health indicators like fetal heart rate patterns and uterine contractions, modern computational tools help identify potential issues that might otherwise go undetected [45].

The use of machine learning and data-driven techniques has transformed how healthcare professionals analyze the vast amount of data generated during pregnancy. These technologies make it possible to detect subtle changes in fetal heart rate and uterine activity, which could indicate distress or abnormalities [46]. With precise pattern recognition and data interpretation, medical experts can make informed decisions about prenatal care and necessary interventions. One of the most significant benefits of advanced analytics in fetal monitoring is the ability to generate real-time alerts for potential complications. By continuously analyzing data from monitoring devices, these systems can quickly detect concerning trends that may pose risks to the mother or baby [47]. Such early warnings enable timely medical intervention, potentially preventing serious health issues and improving maternal and fetal outcomes.

Moreover, integrating computational analytics into fetal monitoring improves the precision and efficiency of medical assessments. With access to extensive datasets, these systems continuously refine their predictive capabilities, leading to more accurate diagnoses. This ongoing learning process allows healthcare providers to offer better evaluations and targeted treatments, ultimately enhancing patient care. Beyond individual patient monitoring, the insights gained through advanced analytics contribute to broader medical research. The large volumes of data collected can help identify patterns in fetal development and health, aiding in the formulation of clinical guidelines and best practices. These findings benefit the entire medical community and drive progress in obstetric care [48].

The incorporation of intelligent analytics in fetal monitoring represents a major advancement in prenatal healthcare. By facilitating detailed data analysis, early risk detection, and improved clinical decision-making, these technologies empower healthcare professionals to provide superior care for both mothers and their babies [49, 50]. As innovations in this field continue to evolve, the integration of sophisticated analytical tools is expected to further enhance maternal and fetal health management.

Conclusion

The incorporation of artificial intelligence (AI) into healthcare analytics is revolutionizing decision-making and operational efficiency. By utilizing AI-driven insights, healthcare institutions can process vast amounts of data, leading to improved patient care and enhanced support for medical professionals. AI-powered predictive analytics allow healthcare providers to examine historical patient data, including electronic health records, genetic profiles, and wearable device information. By identifying trends and predicting potential health risks, medical practitioners can detect diseases earlier, refine treatment plans, and allocate resources more effectively. This proactive approach leads to better patient outcomes, cost savings, and more efficient healthcare services.

AI-driven clinical decision support systems offer real-time recommendations by analyzing patient information in conjunction with medical research. These systems assist in diagnosing complex conditions, suggesting appropriate treatments, and identifying possible drug interactions. With AI integration, clinicians receive precise and timely insights, ultimately improving treatment efficacy and patient safety. Efficient resource management is another significant advantage of AI in healthcare. By analyzing data related to hospital admissions, bed occupancy, and workforce scheduling, AI helps predict demand and optimize resource allocation. This minimizes patient wait times, enhances overall experiences, and streamlines hospital operations. AI also plays a crucial role in detecting fraudulent activities within the healthcare sector. By evaluating extensive datasets related to medical claims, AI can recognize unusual patterns that may indicate fraud. This contributes to reducing financial losses for insurance providers and patients while reinforcing trust in healthcare institutions.

In the pharmaceutical industry, AI accelerates drug discovery and development by analyzing scientific research, clinical trials, and molecular structures. AI-powered analytics can predict the effectiveness of potential drug candidates, refine clinical trial processes, and shorten the time required to bring new treatments to market. This not only reduces costs but also ensures faster access to critical medications.

Additionally, AI enhances pharmacovigilance by assessing data from adverse event reports and other sources. AI algorithms can detect potential safety concerns, prioritize them based on severity, and assist regulatory bodies in conducting further investigations, thereby improving patient safety and regulatory efficiency. In fetal monitoring, AI-powered analytics assist healthcare professionals in analyzing fetal heart rate patterns and uterine activity. By identifying irregularities and potential risks, AI can provide timely alerts, enabling early interventions to safeguard both the mother and the unborn child.

The integration of AI-driven analytics is shaping a more data-oriented and efficient healthcare system. By leveraging AI for predictive modeling, clinical decision support, resource management, fraud detection, drug discovery, pharmacovigilance, and fetal diagnostics, healthcare organizations can enhance care quality and improve patient outcomes. Embracing AI in healthcare analytics represents a step toward a future where data-driven technologies create a more efficient, responsive, and patient-centered medical ecosystem.

References

1. M. Chen and M. Decary, "Artificial intelligence in healthcare: An essential guide for health leaders," *Healthc. Manage. Forum*, vol. 33, no. 1, pp. 10–18, Jan. 2020.
2. Z. Zhang, Y. Genc, D. Wang, M. E. Ahsen, and X. Fan, "Effect of AI explanations on human perceptions of patient-facing AI-powered healthcare systems," *J. Med. Syst.*, vol. 45, no. 6, p. 64, May 2021.
3. B. Kalis, M. Collier, and R. Fu, "10 promising AI applications in health care," *Harvard Business Review*, pp. 2–5, 2018.
4. J. Telo, "AI for enhanced healthcare security: An investigation of anomaly detection, predictive analytics, access control, threat intelligence, and incident response," *JAAHM*, vol. 1, no. 1, pp. 21–37, Jan. 2017.
5. B. Kompa et al., "Correction to: Artificial intelligence based on machine learning in pharmacovigilance: A scoping review," *Drug Saf.*, vol. 46, no. 4, p. 433, Apr. 2023.
6. N. Sambasivan and R. Veeraraghavan, "The deskilling of domain expertise in AI development," Proceedings of the 2022 CHI Conference on Human Factors in Computing Systems, New Orleans, LA, 2022, pp. 1–14.
7. R. R. Dixit and R. P. Schumaker, "A decision tree analysis of opioid and prescription drug interactions leading to death using the FAERS database," IIMA/ICITED Joint, 2018.
8. S. Vyas, M. Shabaz, P. Pandit, L. R. Parvathy, and I. Ofori, "Integration of artificial intelligence and blockchain technology in healthcare and agriculture," *J. Food Qual.*, vol. 2022, May 2022.
9. F. Wang and A. Preininger, "AI in health: State of the art, challenges, and future directions," *Yearb. Med. Inform.*, vol. 28, no. 1, pp. 16–26, Aug. 2019.

10. M. A. Veronin, R. P. Schumaker, R. Dixit, and M. Ogwo, "Irony of the FAERS database: An analysis of data input errors and potential consequences," In *IIMA/ICITED Joint Conference* 2018 (pp. 53–53). International Information Management Association.
11. E. Kim, Y. H. Lee, J. Choi, and B. Yoo, "Machine learning-based prediction of relative regional air volume change from healthy human lung CTs," KSII Transactions, 2023.
12. C. Ye et al., "Prediction of incident hypertension within the next year: Prospective study using statewide electronic health records and machine learning," *J. Med. Internet Res.*, vol. 20, no. 1, p. e22, Jan. 2018.
13. A. Abdullah Alfayez, H. Kunz, and A. Grace Lai, "Predicting the risk of cancer in adults using supervised machine learning: A scoping review," *BMJ Open*, vol. 11, no. 9, p. e047755, Sep. 2021.
14. B. Choi, Y. H. Lee, Y. Kyung, and E. Kim, "Albert with knowledge graph encoder utilizing semantic similarity for commonsense question answering," arXiv preprint arXiv:2211.07065, 2022.
15. A. J. London, "Artificial intelligence and black-box medical decisions: Accuracy versus explainability," *Hastings Cent. Rep.*, vol. 49, no. 1, pp. 15–21, Jan. 2019.
16. F. Pesapane, C. Volonté, M. Codari, and F. Sardanelli, "Artificial intelligence as a medical device in radiology: Ethical and regulatory issues in Europe and the United States," *Insights Imaging*, vol. 9, no. 5, pp. 745–753, Oct. 2018.
17. C. Gilvary, N. Madhukar, J. Elkhader, and O. Elemento, "The missing pieces of Artificial Intelligence in medicine," *Trends Pharmacol. Sci.*, vol. 40, no. 8, pp. 555–564, Aug. 2019.
18. K. P. Gunasekaran, K. Tiwari, and R. Acharya, "Deep learning based auto tuning for database management system," arXiv preprint arXiv:2304.12747, 2023.
19. N. Burkart and M. F. Huber, "A survey on the explainability of supervised machine learning," *JAIR*, vol. 70, pp. 245–317, Jan. 2021.
20. F. Fatani, N. Binhowaimel, M. Alfaisal, M. Aamer, F. Al Asmari, and A. Almuflih, "Machine learning opportunities and challenges in health system management: Review of the existing literature," *Health Informatics J.*, vol. 16, no. 2, pp. 1–18, Dec. 2022.
21. T. Panch, P. Szolovits, and R. Atun, "Artificial intelligence, machine learning and health systems," *J. Glob. Health*, vol. 8, no. 2, p. 020303, Dec. 2018.
22. R. Pugliese, S. Regondi, and R. Marini, "Machine learning-based approach: Global trends, research directions, and regulatory standpoints," *Data Sci. Manag.*, vol. 4, pp. 19–29, Dec. 2021.
23. H. Pallathadka, M. Mustafa, D. T. Sanchez, G. Sekhar Sajja, S. Gour, and M. Naved, "Impact of machine learning on management, healthcare and agriculture," *Mater. Today. Proc.*, vol. 80, pp. 2803–2806, Jan. 2023.
24. P. Uyyala and D. C. Yadav, "The advanced proprietary AI/ML solution as Anti-fraud Tensorlink4cheque (AFTL4C) for Cheque fraud detection," *Int. J. Anal. Exp. Modal. Anal.*, vol. 15, no. 4, pp. 1914–1921, 2023.

25. L. Bote-Curiel, S. Muñoz-Romero, A. Gerrero-Curieses, and J. L. Rojo-Álvarez, "Deep learning and big data in healthcare: A double review for critical beginners," *NATO Adv. Sci. Inst. Ser. E Appl. Sci.*, vol. 9, no. 11, p. 2331, Jun. 2019.
26. H. Abdel-Jaber, D. Devassy, A. Al Salam, L. Hidaytallah, and M. EL-Amir, "A review of deep learning algorithms and their applications in healthcare," *Algorithms*, vol. 15, no. 2, p. 71, Feb. 2022.
27. E. Kim, J. Kim, J. Park, and H. Ko, *TinyML-Based Classification in an ECG Monitoring Embedded System*, CMC, 2023.
28. A. Mohsen, L. P. Tripathi, and K. Mizuguchi, "Deep learning prediction of adverse drug reactions in drug discovery using open TG-GATES and FAERS databases," *Front. Drug Des. Discovery*, vol. 1, pp. 1–9, 2021.
29. A. O. Basile, A. Yahi, and N. P. Tatonetti, "Artificial intelligence for drug toxicity and safety," *Trends Pharmacol. Sci.*, vol. 40, no. 9, pp. 624–635, Sep. 2019.
30. T. Bansal, K. Gunasekaran, and T. Wang, "Diverse distributions of self-supervised tasks for meta-learning in NLP," arXiv preprint arXiv, 2021.
31. M. Munsaka, M. Liu, Y. Xing, and H. Yang, "Leveraging Machine Learning, Natural Language Processing, and Deep Learning in Drug Safety and Pharmacovigilance," in *Data Science, AI, and Machine Learning in Drug Development*, Chapman and Hall, 2022.
32. S. V. G. Subrahmanya et al., "The role of data science in healthcare advancements: Applications, benefits, and future prospects," *Ir. J. Med. Sci.*, vol. 191, no. 4, pp. 1473–1483, Aug. 2022.
33. M. A. Veronin, R. P. Schumaker, and R. R. Dixit, "A systematic approach to cleaning of drug name records data in the FAERS database: A case report," *J. Big. Data*, vol. 7, pp. 1–10, 2020.
34. M. A. Veronin, R. P. Schumaker, R. R. Dixit, and H. Elath, "Opioids and frequency counts in the US food and drug administration adverse event reporting system (FAERS) database: A quantitative view of the epidemic," *Drug Healthc. Patient Saf.*, vol. 11, pp. 65–70, Aug. 2019.
35. G. Trifirò, J. Sultana, and A. Bate, "From big data to smart data for pharmacovigilance: The role of healthcare databases and other emerging sources," *Drug Saf.*, vol. 41, no. 2, pp. 143–149, Feb. 2018.
36. A. Wong, J. M. Plasek, S. P. Montecalvo, and L. Zhou, "Natural language processing and its implications for the future of medication safety: A narrative review of recent advances and challenges," *Pharmacotherapy*, vol. 38, no. 8, pp. 822–841, Aug. 2018.
37. K. P. Gunasekaran, "Leveraging object detection for the identification of lung cancer," arXiv [eess.IV], 25–May–2023.
38. X. Wang, X. Xu, W. Tong, R. Roberts, and Z. Liu, "InferBERT: A transformer-based causal inference framework for enhancing pharmacovigilance," *Front Artif. Intell.*, vol. 4, p. 659622, May 2021.
39. J. Sultana and G. Trifirò, "The potential role of big data in the detection of adverse drug reactions," *Expert Rev. Clin. Pharmacol.*, vol. 13, no. 3, pp. 201–204, Mar. 2020.

40. P. Dhake, R. Dixit, and D. Manson, "Calculating a severity score of an adverse drug event using machine learning on the FAERS database," IIMAJICITED UWS, 2017.
41. A. Bate, K. Hornbuckle, J. Juhaeri, S. P. Motsko, and R. F. Reynolds, "Hypothesis-free signal detection in healthcare databases: Finding its value for pharmacovigilance," *Ther. Adv. Drug Saf.*, vol. 10, p. 2042098619864744, Aug.2019. https://doi.org/10.1177/2042098619864744.
42. E. Kim, J. Lee, H. Jo, K. Na, E. Moon, and G. Gweon, "SHOMY: Detection of small hazardous objects using the you only look once algorithm," KSII Transactions on, 2022.
43. C. L. Ventola, "Big data and pharmacovigilance: data mining for adverse drug events and interactions," *P T*, vol. 43, no. 6, pp. 340–351, Jun. 2018.
44. M. A. Veronin, R. P. Schumaker, and R. Dixit, "The irony of MedWatch and the FAERS database: An assessment of data input errors and potential consequences," *J. Pharm. Technol.*, vol. 36, no. 4, pp. 164–167, Aug. 2020.
45. F. Sabry, T. Eltaras, W. Labda, K. Alzoubi, and Q. Malluhi, "Machine learning for healthcare wearable devices: The big picture," *J. Healthc. Eng.*, vol. 2022, p. 1–25, Apr. 2022.
46. R. R. Dixit, "Predicting Fetal health using cardiotocograms: A machine learning approach," *J. Adv. Anal. Healthc.*, vol. 6, no. 1, pp. 43–57, 2022.
47. R. Chinnaiyan and S. Alex, "Machine learning approaches for early diagnosis and prediction of Fetal abnormalities," 2021 International Conference on Computer Communication and Informatics (ICCCI), 2021, pp. 1–3.
48. M. W. L. Moreira, J. J. P. C. Rodrigues, V. Furtado, C. X. Mavromoustakis, N. Kumar, and I. Woungang, "Fetal birth weight estimation in high-risk pregnancies through machine learning techniques," ICC 2019–2019 IEEE International Conference on Communications (ICC), 2019, pp. 1–6.
49. A. Bertini, R. Salas, S. Chabert, L. Sobrevia, and F. Pardo, "Using machine learning to predict complications in pregnancy: A systematic review," *Front. Bioeng. Biotechnol.*, vol. 9, p. 780389, 2021.
50. A. Zewail, S. Saber, "AI -powered analytics in healthcare: Enhancing decision-making and efficiency," *Int. J. Appl. Health Care Anal.*, vol. 8, no. 5, pp. 1–16, 2023.

Chapter 15

Case Study on Implementing AI for Dynamic Pricing in the Hospitality Industry

Bhupinder Singh

Introduction and Background: AI in Hospitality

AI has consolidated itself as integral in helping organizations improve operational efficiencies, customer journeys, and revenue management practices. Dynamic pricing is one of the most important use cases of AI in hospitality, allowing hotels, resorts, and travel companies to optimize room rates in real time on the basis of different factors that include demand, competitor pricing, seasonality, and customer behavior [1]. Traditionally, hotels used static pricing models, in which prices were established in advance based on expected occupancy rates. But these best-value-allocation models often meant revenue was left on the table during peak periods, and capacity sat idle in off-peak periods. AI creators build dynamic prices utilizing machine learning calculations and massive amounts of information to visualize any reasonable swings in interest and condition cost and ideal income [2].

AI-driven dynamic pricing enhances profitability, efficiency, and competitiveness. AI systems deliver improved pricing recommendations, utilizing historical data, real-time booking trends, weather conditions, and even global economic indicators over traditional revenue management

 DOI: 10.4324/9781003606956-15

formats. Furthermore, it also enables hospitality enterprises to maintain a competitive edge in a rapidly changing market landscape through customized pricing, where rates are adjusted based on customer taste preferences, loyalty programs, and booking patterns [3].

AI improves more than just pricing in the hospitality industry playing the roles in customer service, predictive maintenance, automated check-ins, and chatbots for customer interaction. AI-enabled recommendation engines are already employed by platforms such as Airbnb and Expedia to recommend tailored accommodations and pricing to help improve guest experience. As AI progresses, its capabilities will enable even more nuanced applications in devising and implementing pricing strategies, making AI indispensable in any modern revenue management toolkit [4].

Development of Dynamic Pricing Strategies

Dynamic pricing is not new; this approach has been developing for more than 20 years, first in airline ticketing and e-commerce before being adopted in hospitality. In the past, hotel pricing was a manual approach where revenue managers would base rates on seasonality, occupancy rates, past demand trends [5]. The second wave of dynamic pricing in hospitality began with yield management systems (YMS), introduced in the airline industry in 1980. Hotels quickly followed suit, deploying emerging computerized systems to revise prices based on predicted demand. But these systems were also quite rigid and not nearly as flexible as modern AI-powered solutions.

The second wave started with the rise of online travel agencies (OTAs) like Booking.com and Expedia in the early 2000s. These platforms allowed hotels to adopt dynamic pricing according to real-time market trends. Hoteliers were using rule-based pricing models to set different price points depending on occupancy thresholds, competitor rates, and advance booking windows. This third wave of dynamic pricing powered by AI and machine learning uses algorithms that reactively adjust prices in real time based on many internal and external factors. AI-driven revenue management solutions employ big data, and predictive analytics, along with automation, to make pricing decisions more efficient [6]. These systems take into account everything from customer behavior to competitive intelligence to macroeconomic indicators to decide the right price to set at any given moment in time. AI-powered pricing is far more accurate and agile than legacy solutions, so hoteliers can maximize revenue without alienating customers. Today, you have evolved both models from static pricing based to highly intelligent, automated pricing models, representing a paradigm shift at the hospitality sector where real-time data processing and AI-driven insights dictate competitive advantages [7].

Understanding Dynamic Pricing in the Hospitality Industry

Dynamic Pricing: Dynamic pricing allows organizations to adapt to changes in demand. This methodology helps hotels, resort properties, and vacation rental homes contribute to revenue management by being able to sell at various prices for various customers at various times in the hospitality industry. Dynamic pricing, on the one hand, uses real-time data analytics and machine learning algorithms to set the best price for things like hotel rooms, packages, or services. Static pricing, on the other hand, has rates that don't change much over long periods of time. Dynamic pricing relies heavily on four main principles, which are Demand-Based Pricing, Segmentation, Elasticity, and Price Differentiation [8].

Demand-Based Pricing: Prices vary with fluctuating supply and demand. In peak travel seasons, there's a demand, and prices go up; in the controlled off-peak periods, we have lower prices that stimulate the bookings.

Segmentation: Different segments of customers demonstrate varying sensitivities to pricing. Business travelers are often last-minute bookers and can pay a premium, while leisure bookers are typically more price-sensitive and plan in advance.

Elasticity: Determining the price can depend on how sensitive customers are to price changes. When demand is consistent regardless of price increases, hotels have the opportunity to take advantage of premium pricing [9]. On the upside, competitive rates keep occupancy up if customers are very sensitive to price.

Price Differentiation: Hotel competitors closely monitor each other and position their pricing competitively to attract more guests while keeping a healthy margin.

Hospitality dynamic pricing models incorporate booking patterns, cancellation probabilities, customer loyalty, and macroeconomic trends as well. AI-led revenue management systems run through multiple data points, among many other factors, like weather conditions, major events and online search trends, to make pricing decisions. Hoteliers best maximize revenue while customers receive pricing tailored to them based on propensity to book, price sensitivity, and overall preferences [10].

Price Models in Hospitality: Historical Approaches

Fixed Pricing Models (Pre-1980s)

Traditionally, hotels followed fixed pricing models, setting room rates for a season or an entire year. Not only was this method inflexible, but it also frequently cost revenue by overcharging in a low demand cycle or undercharging in a high demand cycle. Accurate booking and pricing were set based on the median of past bookings, but not taking into account the current demand, leading to revenue losses due to not maximizing potential users [11].

Yield Management (1980s–2000s)

Inspired by the implementation of yield management systems in the airline industry in the 1980s, hospitality soon began implementing similar models. With the advent of yield management systems (YMS), hotels started to segment their inventory so they could charge different prices depending on the booking window, cancellation policy or guest type. These systems were largely reliant on fixed rules and human input, which limited their ability to adapt to dynamic market changes.

Online Travel Agency (OTAs) and Price Optimization (2000s–2010s)

OTAs such as Expedia, Booking.com, and Airbnb upended hotel pricing. Real-time data access enabled hotels to change their rates dynamically according to occupancy numbers, competitor prices, and demand forecasts [12]. While rule-based pricing thus enhanced revenue management, it failed to deliver the required agility to adapt to changing market conditions.

2010s–Present: AI-Powered Dynamic Pricing

Contemporary hotels use AI and machine learning to optimize for demand variations, evolve prices in real-time, and personalize deals. From dynamic pricing on e-commerce platforms to yield management in airlines, companies are now using machine learning-based models to analyze vast amounts of data, be it social media trends, economic indicators, or customer preferences, in order to devise optimal pricing strategies [13]. Today's data-driven dynamic pricing systems test, learn, and adjust automatically, in real time, and without human interference, resulting in enhanced profitability and improved customer satisfaction.

Modern Pricing Models: The Role of Technology

Increases in new technologies enhance the dynamic pricing strategies in the hospitality industry. Today they are equipped with advanced AI-powered tools, data analytics, and cloud-based platforms to automate pricing decisions, analyze real-time data, and predict changes in demand as never before [14]. Machine learning (ML) models analyze data continuously and learn from patterns unlike the traditional rule-based systems, therefore refining the pricing strategies over time. Proactive pricing, RMS (revenue management system) driven by ML customize pricing for each client in real time, depending on changes in supply and demand in real time; booking pace (how fast rooms are being booked); customer segmentation and willingness to pay.

Dynamic pricing tools like Duetto, Revinate, and Pace Revenue leverage ML models to automate pricing decisions making them less reliant on manual intervention and improving pricing precision. However, modern hotels have cloud-based RMS platforms, which interfaces directly with property management systems (PMS), customer relationship management (CRM) tools, and OTAs. These platforms offer up-to-the-minute updates on market trends, customer interests, and revenue trends. With algorithms analyzing customer profiles, loyalty programs, and browsing history, AI takes personalized dynamic pricing to the next level, automatically adapting room rates in real time [15]. New technologies' applications like blockchain help to generate more transparent and tamper-proof pricing models. In addition, smart contracts based on the blockchain can create a transparent and unchangeable database of price changes, ensuring fair pricing among hotels and OTAs alike and mitigating any price discrepancies.

Dynamic pricing as a strategy is easily swept up in the changes technology brings to the pricing landscape, turning it into a near-half-life, AI-optimized, real-time strategy. The implementation of such cutting-edge technologies including big data and machine learning, cloud-based RMS, and AI-based personalization at hotels and restaurants ensures greater revenue potential, improved operational efficiency, and enhanced guest satisfaction [16]. Artificial intelligence, too, is an inscrutable yet powerful beast in the service of dynamic pricing; it can quickly fine-tune decisions so that any hotel can respond in real time to changes in local demand or shifts in reservations.

AI-Driven Dynamic Pricing with Machine Learning Algorithms for Pricing Optimization

Dynamic pricing, also known as demand-based pricing, is a variable pricing strategy where prices are adjusted in real time according to supply and demand, competitor pricing, seasonality, customer information and preferences, etc. Within the hospitality sector, hotels and resorts are increasingly using dynamic pricing to optimize revenue and broaden their occupancy rates while creating a better experience for customers [17]. Still, because traditional pricing models depend on manual iterations and past information, such limitations hinder them from adjusting to real-time margins in a prompt manner.

Artificial intelligence (AI) and machine learning (ML) have now been integrated into the dynamic pricing field, making it more dynamic, real-time, and data-driven than ever. Dynamic pricing models powered by AI leverage complex algorithms, predictive analytics, and automation to determine and adjust prices in real time [18]. These models ingest massive volumes of historical and real-time data, discerning intricate patterns and interconnections that no human analyst could even hope to discern in a reasonable timeframe. In the hospitality sector, there are a number of widely used machine learning algorithms that are useful for optimizing pricing strategies. These consist of supervised learning, unsupervised learning, and reinforcement learning models.

Supervised learning algorithms train on historical pricing data and known lever demand, to be able to predict future pricing trends. Such models assist hotels in predicting room prices under various market scenarios. That is where big data comes in; they can leverage multiple weak predictors into a strong model together to increase accuracy on the pricing [19]. With the aid of machine learning algorithms, hotels can utilize dynamic pricing based on data collected, helping hotels optimize overall revenue, competitiveness, and, eventually, guest satisfaction. AI-powered pricing systems analyze large volumes of data using supervised, unsupervised, and reinforced learning models and adjust pricing in real time to maximize profitability. AI technology is evolving at an unprecedented pace and hotels that adopt AI-powered dynamic pricing strategies will be better positioned to stay ahead of this increasingly competitive sector, ultimately leading to sustainable revenue growth and customer-centric pricing models [20].

Case Study: Implementing AI for Dynamic Pricing in a Hospitality Business

One of the most important decisions in the hospitality industry is to select a suitable pricing strategy, as it directly affects the firm's performance and success. Companies operating in this sector, whether luxury hotels, mid-range stays, budget hotels, holiday rentals or online travel agencies (OTAs), depend on sophisticated pricing strategies to remain competitive and maximize revenue. Traditional pricing decisions were carried out manually, relying on past data, intuition, and rigid seasonal pricing models. But with the introduction of artificial intelligence (AI) and machine learning (ML), businesses were offered the opportunity to adopt AI-powered dynamic pricing strategies that allowed for data-driven decisions to be made in real time to adjust pricing [21]. Dynamic pricing solutions powered by AI enable demand-based hospitality business response along with changes in competitor pricing, booking patterns, and external market conditions including (but not limited to) local events, holidays, economic trends, and weather changes. This evolution enables companies to price better based on market demand, maximize revenue per available room (RevPAR), and balance occupancy without human interaction 24/7. From luxury resorts trying to ensure exclusivity while maximizing pricing to OTAs whose recommendation engines are AI-backed, to business hotels that want a flexible pricing model, and budget hotels that wish to scale their availability for surge demand to OTAs whose recommendation engines are AI backed, AI-driven pricing models are prevalent across most hospitality player types [22].

The model's initial stage consists of extensive data collection and integration from several sources: internal business data as historical booking patterns, guest demographics, past revenue trends, external market data for competitor pricing, local demand trends, real-time economic indicators and third-party data (social media trends, traveler sentiment analysis, related to travel destinations' search engine activity, etc.). These data are aggregated and processed constantly by AI systems to ensure that pricing decisions are based on the most relevant and updated market conditions [23]. The subsequent data processing encompasses machine learning algorithms that churn through the data to provide insights into demand forecasting, customer booking patterns, and price elasticity. Advanced predictive analytics algorithms and deep learning models find patterns and correlations in booking behaviors, seasonal peak demand variations, and competitors' pricing strategies. Reinforcement learning is employed by AI systems to iteratively improve pricing models through real-time feedback, enabling dynamic price adjustments based on the observed effectiveness of prior pricing strategies.

Adjusted prices are then automatically updated in real time on various distribution channels directly managed by the hotel (direct booking) as well as indirect ones such as OTAs, corporate travel booking systems, and Global Distribution System (GDS). The AI maintains the rate parity in all channels but allows businesses to deliver specific pricing in the segmentation of customers [24]. Business travelers, for example, may be charged differently than leisure travelers; if they are loyal customers or long-term guests, they could get tailored discounts according to their booking pattern. For example, the AI system also oversees real-time market shifts such as competitors' rates, sudden spikes in demand because of events or holidays, and unexpected cancellations, changing prices in seconds to seize revenue opportunities [25].

AI-powered dashboards show real-time information on key performance indicators (KPIs) like occupancy rates and RevPAR, average daily rate (ADR), and booking trends. This is followed by tracking pricing effectiveness and areas of improvement are analyzed and, based on that, data-driven strategies are implemented through dashboards. So, factors like pricing generally come under the transport policies, which are scribed with the help of A/B testing in AI systems, where the impact of various pricing strategies is evaluated systematically and the return on revenue generation, booking behavior by customers, and percentage of occupancies along with sellouts is measured [26]. Businesses can also configure rules so AI has bounds on how to operate, such as minimum and maximum pricing thresholds, blackout dates, and promotional periods.

There is also no minimum margin from which profit is derived or a specific product line; rather, it reduces human interaction in pricing and ultimately automates that process, which both optimizes revenue management and enhances operational performance. Businesses can optimize pricing in real time and stay ahead of the game in the market by automating the entire pricing process, eliminating human errors, and allowing for complex calculations. AI-based dynamic pricing enhances the customer experience by providing fair and transparent pricing, personalized discounts, and availability optimization according to demand patterns [27]. AI-driven pricing solutions empower hospitality businesses to leverage technology to navigate these challenges effectively, ensuring sustained revenue growth and long-term success. Today, AI-powered dynamic pricing has gone from nice-to-have to need-to-have for any business looking to stay relevant in the fast-moving hospitality space.

Tools and Technologies Used in Pricing Optimization: Real-Time Data Sources and Price Fluctuation Models

The hospitality sector is witnessing practical advantages from artificial intelligence (AI) and machine learning (ML)-based pricing strategy optimization. Dynamic pricing models powered by AI use a combination of advanced tools, sophisticated algorithms, and data sources in real time to set clear and automatic pricing decisions [28]. These solutions leverage algorithms to interpret demand trends, competitor pricing strategies, market fluctuations, and customer behavior patterns, and others to deliver optimal pricing. Revenue management system (RMS) is one of the most powerful pricing optimization tools available; the software uses AI-powered analytics to consider historical booking patterns, occupancy levels, and demand forecasts. Major RMS software solutions like Duetto, IDeaS, RateGain, and Cloudbeds connect to hotel PMS systems and channel managers to automatically refresh prices on countless OTAs and direct booking sites. Predictive analytics and demand forecasting algorithms give users the most competitive prices, while keeping profit margins intact and demonstrating the same rates on all distribution channels [29].

Dynamic pricing engines are another vital technology in AI-led pricing optimization. These algorithms use machine learning to keep improving their pricing strategies as they receive real-time data inputs.

Dynamic pricing engines observe things like competitor price movements, seasonality, local happenings, customer reviews, and even the weather, to adjust room rates. For example, if there is a big conference or festival in a city, AI-powered pricing engines can sense the increase in demand and adjust room prices ahead of the competition [30]. If cancellations are high and inventory suddenly becomes excess, the system can manage to discount calmly to drive last-minute bookings. With booking engines, cloud-based dynamic pricing solutions ensure they have the functionality to connect seamlessly, allowing for automatic price updates at the point of sale in real time without any manual input.

AI chatbots and AI-enabled customer relationship management (CRM) systems also help businesses gain valuable customer insights for pricing optimization. Artificial intelligence chatbots can also use natural language processing (NLP) to moderate consumer queries, preferences, and booking habits to execute individualized pricing techniques. For instance, if returning customers often make hotel reservations for certain travel seasons, AI can identify this behavior and provide targeted discounts or loyalty incentives [31]. And sentiment analysis tools also monitor the tone of online reviews, social media mentions, and travel forums to gauge brand reputation and demand perception, which feeds back into dynamic pricing algorithms.

AI-driven dynamic pricing models are fueled by real-time data sources that keep pricing strategies agile in the face of market fluctuations. The main data sources consist of GDSi, web scraping tools, API integrations with OTAs and meta-search engines (e.g., Google Hotels, TripAdvisor, Kayak). Real-time data about competitor pricing, traveler search intent, and rate conversion is available on these platforms. These extensive datasets are analyzed by machine learning algorithms to spot pricing trends and demand fluctuations, allowing hotels and travel agencies to optimize their pricing strategies [32]. The macroeconomic indicators, including currency rates, fuel prices, and any geopolitical events affecting demand for travel, feed into AI models to create cost-effective pricing in response to changes in the economic environment.

Predictive analytics, reinforcement learning, and regression analysis techniques used in AI-based dynamic pricing systems are used to build price fluctuation models. With predictive analytics, booking history and external factors are utilized as inputs to predict demand structures, enabling businesses to more accurately predict peak and off-peak seasons. Reinforcement learning algorithms experiment with a range of pricing strategies, using insights from past pricing effectiveness to enhance rate alterations in the future. For instance, AI may learn from historical data that rooms with a sea-view suite or ones with executive lounge access see higher demand over weekends and will set prices accordingly [33]. Regression analysis also allows for more advanced models of price movement by correlating pricing activity with demand (often measured in terms of occupancy), in order to ensure that pricing activity does not stray too far from revenue targets.

Incorporating real-time occupancy tracking, dynamic pricing models ensure that hotels derive the maximum profitability while controlling inventory. AI-based pricing models are occupancy based and adjust room rates dynamically according to real time or predicted occupancy rates. A hotel may have lower occupancy during off-peak seasons, so AI algorithms may dynamically provide flash discounts or bundled package deals to entice bookings. In contrast, during peak demand periods, AI models limit promotional discounts and adopt surge pricing to take advantage of peak demand [34]. Customers look also for advanced real-time dashboards that allow revenue managers to stay on top of pricing performance and adjust AI-driven strategies with a precision thumb whenever needed.

Blockchain technology is also beginning to play a role in AI-driven pricing optimization by ensuring data transparency, security, and reliability. Blockchain enables smart contracts to automate pricing agreements between customers, hotels, and OTAs, minimizing the risk of price manipulation or unauthorized rate changes. Such contracts can also cross-validate real-time demand changes across various platforms and guarantee consistency in dynamic pricing updates. By leveraging AI, machine learning, and real-time data sources for pricing optimization, hospitality businesses can gain a substantial competitive edge [35]. Automation contributes to efficient pricing practices. It comes with automated revenue management systems, dynamic pricing engines, customer sentiment analysis, and blockchain-enabled transparency that can take your data-driven pricing decisions a step further in terms of profitability and customer satisfaction. This AI-driven dynamic pricing empowers hospitality businesses to stay aligned with market trends and forge a forward-thinking, flexible pricing strategy that meets the demands of the digital age.

Impact of AI-Based Pricing Strategies on Hospitality Businesses

AI in pricing has become a game changer in the hospitality industry, enabling hotels, resorts, and travel agencies to generate the optimal revenue and create a fruitful customer experience like never before.

What are Dynamic Pricing Strategies?

AI-based dynamic pricing strategies rely on machine learning, big data analytics, and predictive analytic techniques to execute dynamic pricing intelligence as decision-making in real time to maximize profitability, operational efficiency, or create personalized experiences for guests [36]. Strategic approaches take on significance in their effects across various dimensions of the hospitality sector: revenue generation, competitive positioning, customer satisfaction, operational sustainability, and so forth.

AI-based pricing strategies have a significant impact on revenue management [37]. Static pricing models tend to be unable to account for rapidly changing demand, losing sales after classes and events have been overbooked or losing revenue due to underpricing. Dynamic pricing algorithms powered by artificial intelligence assess market conditions, competition pricing, booking patterns, and historical data to optimize prices in real time [38]. By understanding demand elasticity, hospitality businesses can adjust pricing to maximize revenue during peak periods while attracting value-sensitive travelers during off-peak seasons. The industry has responded by automating pricing decisions and underpinning advanced yield management strategies that line up occupancy level and profit margin through special artificial intelligence (AI)-driven revenue management systems (RMS) like IDeaS, Duetto, and RateGain.

The more significant benefit of AI pricing is improving competitive positioning in a saturated market. The hospitality industry is highly competitive, where hotels, vacation rentals, and online travel agencies (OTAs) are constantly tweaking prices to land customers. AI-based pricing tools allow companies to track competitor pricing in real time and promptly update it to stay one step ahead. For example, if a competitor hotel drops the room prices due to low occupancy, such a system can easily identify the change around the market and recommend a fit pricing which keeps the business moving while making a reasonable quota. AI models can also anticipate demand fluctuations during large events, holidays, or economic changes: hotels can then proactively optimize pricing ahead of competitors responding to the same market opportunities.

Pricing strategies based on AI have also a considerable influence on customer experience and satisfaction [39]. While traditional pricing mechanisms are often based on overall price movements, AI-driven systems have the capability to dynamically personalize pricing for each customer, based on variables like the customers' preferences, reservation history, and loyalty status [40]. Based on past customer interactions, stay patterns and spending behavior, machine learning algorithms analyze and provide customized discounts, dynamic loyalty pricing, personalized promotions etc. Returning guests who often book premium suites might be offered deals tailored to a more historical behavior, thereby driving brand loyalty and boosting direct bookings. AI-driven chatbots and virtual assistants enhance customer engagement by delivering real-time pricing data, personalized suggestions, and immediate booking support.

AI-powered pricing strategies drive efficiency and cost reduction from an operational standpoint. Automated pricing also saves revenue managers and hotel staff from the hassle of manual rate adjustments. AI models can ingest billions of data points in seconds and make pricing decisions in moments that can take analysts hours to calculate. Additionally, this level of automation reduces errors, promotes uniform pricing across various booking platforms, and prevents revenue loss caused by outdated pricing. AI-based demand forecasting aids hospitality organizations in optimizing staffing, inventory control, and resource distribution, resulting in considerable cost savings. For instance, if an AI model predicts a spike in bookings for an upcoming festival, hotels can schedule staff, housekeeping, and amenities as necessary, leading to better service quality and guest satisfaction.

So AI-based pricing strategies not only enable businesses to grow but also promote sustainability. Using real-time data to inform decisions enables hotels to optimize their room utilization, minimize waste of electricity, and facilitate sustainable pricing models that promote green travel. Some AI-powered platforms incorporate real-time carbon footprint tracking into pricing strategies so that hotels can offer guests incentives on eco-friendly travel options, the route between return trips to the same destination or selecting energy-efficient accommodations, such as opting to forgo daily housekeeping [41]. It is also in line with global sustainability initiatives while adding to the reputation value of hospitality businesses that adopt environment friendly practices.

While there are advantages, there are also challenges and ethical implications for AI-based pricing strategies. One of the major concerns is price discrimination and fairness. Dynamic pricing maximizes revenue, in theory, but can lead to widespread price differences for customers booking at different time stamps. If consumers feel they're getting price gouged because prices are driven by AI, then retailers may find themselves with bad press. To tackle this issue, hospitality companies need to provide a clear pricing policy and a fairness-centric AI model, deterring any predatory pricing [42]. The reliance on AI and automation can lead to a feeling of coldness in the hospitality industry. AI best improves pricing efficiency, but should be an aid to, rather than a substitutive, human activity. While personalized hotel experiences continue to be important for many guests, automating pricing processes too much could make it hard to adjust for one-off guest requirements [43]. As a result, it calls for a hybrid approach that pairs artificial intelligence-powered insights with human know-how in order to uphold the hospitality industry's bedrock of service above all [44].

The future prospects of AI-based pricing strategies in hospitality will keep on being perfected along with the increasing developments in machine learning, blockchain technology, and AI-powered sentiment analysis. The AI model of the future will be predicting customer emotions and sentiment from online reviews, social media mentions, and direct feedback as well as optimizing pricing, allowing businesses to modify prices according to brand reputation and guest satisfaction perception. With the integration of blockchain in dynamic pricing, this will create price transparency and eliminate any rate disparities on various booking websites, giving price fairness to all customers.

AI-based pricing strategies have revolutionized the hospitality industry, delivering significant revenue boosts, new levels of customer satisfaction, and operational efficiencies. Using data from predictive analytics, machine learning, and real-time sources, hospitality businesses can optimize pricing to meet customer and profitability expectations. Mitigating any ethical concerns, ensuring transparency, and combining AI with human decision-making will be essential for enabling the sustainable adoption of AI-driven pricing strategies. With the advancement of AI, hospitality businesses can leverage these technologies to adopt intelligent pricing models, allowing them to stay ahead of the curve in the rapidly evolving tech landscape.

Challenges and Ethical Considerations in AI-Based Pricing

AI-powered dynamic pricing has revolutionized the hospitality sector by facilitating real-time pricing adjustments in response to demand changes, competitor prices, and customer behavior. Nevertheless, even with all its benefits, AI dynamic pricing can raise some challenges along with ethical questions that need to be addressed to provide fair pricing to all customers while maintaining customer trust and regulatory compliance. Like any powerful tool, price discrimination models built on artificial intelligence (AI) may also backfire, as they raise significant issues related to algorithmic bias, price discrimination, data privacy, and regulatory compliance, as well as an alleged lack of fairness when AI-based analysis results in price discrimination [45]. As AI-driven pricing tools take center stage in revenue management practices, it has become imperative for businesses to address these challenges to enforce transparency, ethics, and customer-centricity in pricing models.

The big challenge of AI-based pricing is algorithmic bias and fairness. Machine learning algorithms and AI models are trained on historical datasets, and if these datasets include biases, be they socioeconomic, demographic, geographical, or otherwise, the pricing algorithms can inadvertently perpetuate those biases. For example, machine learning algorithms may charge higher prices to customers from certain geographical areas, income levels, or browsing behaviors, creating unintended discriminatory outcomes. This phenomenon can raise an ethical dilemma when AI-powered systems bias services against segments of customers, potentially making these services less accessible to economically disadvantaged groups. In order to combat this, companies should ensure that artificial intelligence models are trained on datasets that are diverse and free from bias, and review those models regularly to ensure they are fair and do not enable discriminatory pricing [46].

The other significant ethical issue is about price discrimination and transparency. Dynamic pricing fueled by AI can cause dynamic pricing to fluctuate in response to demand, customer data, and market trends even creating wide price swings in instances. A hotel room price, for instance, might jump dramatically in minutes if there's a sudden increase in online searches or if a rival changes its price. Although businesses claim this kind of pricing increases revenue, consumers see this as inequitable. If a customer notices variation in the price of the same service at different times, they might somehow feel manipulated, which will lead to dissatisfaction and trouble to the business reputation [47]. To combat this, hospitality businesses should adopt a clear and transparent pricing policy that helps customers understand how their prices are determined and what kind of fair bookings are available to them. Other companies have price caps and fair-pricing algorithms to avoid extreme spikes that may turn off consumers.

The crucial challenge in AI-driven pricing strategies is data privacy and security. AI analytics leverages this huge volume of personal data to make pricing decisions, such as customer booking history, online behavior, and location tracking. Although this data is critical for fine-tuning pricing and enhancing the customer experience, there are also serious implications surrounding data security and adherence to international privacy laws, like the General Data Protection Regulation (GDPR) and California Consumer Privacy Act (CCPA). Improper access to customer data or abuse of personal data could result in legal action and a loss of consumer trust [48]. The hospitality industry is also at risk of a legal backlash if personal data is mishandled or breached; losing customers and trust in their business is far too high a price to pay. It is equally important for data collection practices to be transparent, and for customers to have a say in how their data is used to inform price-setting decisions.

AI-based pricing in the hospitality sector presents additional challenges in terms of regulation. Although AI-based pricing has gained acceptance in many areas, some governments have taken action to curb dynamic pricing, especially within critical industries, like transportation and lodging. Regulators are increasingly focusing on AI's role in price setting when it leads to excessive price spikes in times of high demand, like natural disasters, pandemics, or large public events. Some areas are enacting anti-price gouging laws to curb price gouging, and hospitality businesses need to ensure their AI pricing models comply. Legal scrutiny of competitive practices involving AI is also increasing, with some warning that algorithmic pricing could enable automated collusion, where AI systems across competing companies learn to set pricing so as to limit competition and raise prices artificially.

The other big challenge is the conversation of human attention and agency. Although AI pricing streamlines processes, over-reliance on automated pricing may result in inflexible pricing allocations that do not factor in extenuating customer circumstances or unanticipated market dynamics. For instance, a pricing algorithm might raise rates drastically amid a sudden surge of demand yet fail to account for humanitarian or ethical factors, like the necessity of affordable lodging at times of emergency.

The hospitality industry is based on customer service and human relationship and too much automation can lead to a loss of flexibility in a pricing decision that human revenue managers can provide. To prevent this from happening, companies need to leverage a mixed pricing approach, in which AI-generating recommendations and final price point decisions are in correlation with how humans perceive it with an ethical viewpoint.

The important factor to consider is the impact of AI-driven pricing on customer trusts and loyalty. When customers believe AI pricing is opaque or unfair, the experience can erode trust in the brand. Research into consumer behavior has found that people tend to remain loyal to businesses they feel are honest in their pricing, as long as that pricing remains consistent, even when they could receive cheaper prices elsewhere with a dynamic approach. The aim of AI pricing cannot be to make as much top line as possible before customer trust is lost. Therefore, by adopting an ethical AI, responsible pricing paradigm with principles such as ensuring pricing fairness, explaining price change rationales to customers and offering alternatives, including discounts based on loyalty functions results in utilizing a sustainable AI to solve a short to mid-term problem while creating a long-term relationship with the customer [49].

There are also ethical issues surrounding the use of AI for surge pricing and in crisis situations. To some, it could appear like exploitation due to stinger massive price increases during times of higher demand. For instance, during a natural disaster or a health crisis, AI-based algorithms may hike hotel prices because of higher demand, resulting in charges of price gouging. Ethical surge pricing frameworks that did not impose caps and humane exceptions to prevent price gouging were missing. A few have rolled out "ethical AI pricing models" such as pausing or limiting price increases during crises to preserve goodwill and social responsibility.

The technological requirements of deploying AI-based pricing solutions should not be ignored. AI also requires a considerable amount of infrastructure, access to the vast datasets, and constant monitoring to ensure accuracy and fairness. This all begs the question: Can smaller hospitality companies, without large digital teams, really access this kind of technology to be competitive? And integrating AI pricing with existing reservation systems and third-party booking platforms can be complex. However, companies need to devote more resources to retraining, AI governance frameworks, and ethical auditing mechanisms to ensure that AI-based pricing remains not only profitable but also ethical. AI-based pricing strategies have tremendous value for the hospitality industry, yet they also have major challenges and ethical questions. Ridding the society of algorithmic bias, ensuring pricing fairness, transparency, protecting the data of consumers along with regulatory compliance are critical to ensuring the sustainable implementation of AI in the pricing models [50]. AI-driven pricing presents an opportunity for hospitality businesses to innovate their pricing models, using data to inform pricing strategies that maximize revenue and achieve meaningful outcomes. This includes not only leveraging cutting-edge AI technology but also understanding its limits and the ethical implications that accompany its adoption.

Conclusion and Future Prospects of AI in Hospitality Pricing

Data-driven decision-making, coupled with revenue management and revenue optimization techniques, have touted AI and its importance across industries, including hospitality, travel, and tourism. Dynamic pricing through AI helps businesses assess real-time data to anticipate demand changes and pricing so they can be more profitable and efficient. However, with the advancements of AI pricing systems, businesses need to navigate through a set of challenges like algorithmic biases, ethical implications, data privacy issues, and compliance with regulations. Although its potential in transforming pricing strategies has already been proven, AI's success in the long run within the hospitality industry will rely on responsible implementation, constant innovation, and a customer-centric strategy.

Real-time price optimization is one of the most compelling advantages of AI-driven pricing. Unlike conventional pricing strategies that depend on pre-established rates or periodic adjustments, AI facilitates dynamic, data-driven decision-making that can adapt in real time to changes in market conditions. Trained on data collected all the way up to October 2023, for instance, these machine learning algorithms examine a wide range of data points, such as occupancy rates, competitor pricing, seasonal trends, and consumer behavior to determine the optimal price at which to sell rooms in order to maximize revenue. It decreases the dependency on human decision-making, decreases the risk of pricing errors, and enables companies to scale up when responding to demand shifts. Moreover, AI-driven predictive analytics improves demand forecasting, allowing businesses to plan their inventory, staffing, and marketing strategies more efficiently.

Despite the benefits, challenges, and ethical considerations remain critical aspects of AI-driven pricing. Algorithmic bias, opacity of pricing, and over-reliance on automation can produce adverse customer perceptions and erode trust. However, if implemented in a manner more akin to exploitation, the practice of price discrimination may leave some segments of the population with a bad taste, and would likely have to contend with both governmental and social opposition to their business practices. Furthermore, as AI systems need access to massive quantities of consumer data to refine pricing decisions, issues of privacy are becoming more and more central. Mapping the data required of the company can also make it necessary to create data governance policies to prevent regulatory issues in the future, preventing the use of overlaps according to the General Data Protection Regulation (GDPR) and California Consumer Privacy Act (CCPA) regulations.

So, going forward, an AI-powered hospitality pricing future can be anticipated that focuses more on fairness, transparency, and personalization for customers. However, you will start to see that the businesses that conduct themselves ethically when deploying AI to update their pricing strategies, and clear communication in regard to their practices, quickly gain the upper hand over competitors to generate brand loyalty. Finally, hybrid AI-human pricing models can help ensure that any AI-driven pricing recommendations are balanced with human oversight to avoid potential unethical practices. The important trend shaping the future of AI-based pricing is the inclusion of advanced AI techniques like deep learning, reinforcement learning, and explainable AI (XAI). Better pricing accuracy, decision-making, and interpretability will strengthen public confidence in AI-based pricing models by ensuring that customers and regulators understand how a model derived its pricing formula. In addition, AI will also evolve to fit in with the overarching trends across the globe, such as the increasing move toward sustainable tourism, personalization in pricing structures, and the increasing demand for adjusting value-driven pricing structures. AI-assisted pricing will shape the future of the hospitality industry like no other. Despite the challenges ahead, businesses implementing ethical, customer-focused AI strategies will ultimately have the best opportunity to unlock its true potential, thus securing sustained growth, competitiveness, and consumer trust in an increasingly digital world.

References

1. Dwivedi, Y. K., Pandey, N., Currie, W., & Micu, A. (2024). Leveraging ChatGPT and other generative artificial intelligence (AI)-based applications in the hospitality and tourism industry: practices, challenges and research agenda. *International Journal of Contemporary Hospitality Management, 36*(1), 1–12.
2. Gupta, R., Nair, K., Mishra, M., Ibrahim, B., & Bhardwaj, S. (2024). Adoption and impacts of generative artificial intelligence: Theoretical underpinnings and research agenda. *International Journal of Information Management Data Insights, 4*(1), 100232.
3. Khan, U., & Khan, K. A. (2024). Generative artificial intelligence (GAI) in hospitality and tourism marketing: Perceptions, risks, benefits, and policy implications. *Journal of Global Hospitality and Tourism, 3*(1), 269–284.
4. Wang, P. Q. (2024). Personalizing guest experience with generative AI in the hotel industry: there's more to it than meets a Kiwi's eye. *Current Issues in Tourism, 28*(4), 527–544.
5. Alyasiri, O. M., Selvaraj, K., Younis, H. A., Sahib, T. M., Almasoodi, M. F., & Hayder, I. M. (2024). A survey on the potential of artificial intelligence tools in tourism information services. *Babylonian Journal of Artificial Intelligence, 2024*, 1–8.
6. Mogaji, E., Viglia, G., Srivastava, P., & Dwivedi, Y. K. (2024). Is it the end of the technology acceptance model in the era of generative artificial intelligence?. *International Journal of Contemporary Hospitality Management, 36*(10), 3324–3339.
7. Gursoy, D., & Cai, R. (2024). Artificial intelligence: An overview of research trends and future directions. *International Journal of Contemporary Hospitality Management, 37*(1), 1–17.
8. Kannan, R. (2024). Revolutionizing the tourism industry through artificial intelligence: a comprehensive review of AI integration, impact on customer experience, operational efficiency, and future trends. *International Journal for Multidimensional Research Perspectives, 2*(2), 01–14.
9. Singha, R., & Singha, S. (2024). Building capabilities and workforce for metaverse-driven retail formats. In *Creator's economy in metaverse platforms: Empowering stakeholders through omnichannel approach* (pp. 111–131). IGI Global.
10. Abrokwah-Larbi, K. (2024). Transforming metaverse marketing into strategic agility in SMEs through mediating roles of IMT and CI: Theoretical framework and research propositions. *Journal of Contemporary Marketing Science, 7*(1), 56–83.
11. Caboni, F., & Hagberg, J. (2019). Augmented reality in retailing: A review of features, applications and value. *International Journal of Retail & Distribution Management, 47*(11), 1125–1140.

12. Yang, L., Kumar, R., Kaur, R., Babbar, A., Makhanshahi, G. S., Singh, A., ... Alawadi, A. H. (2024). Exploring the role of computer vision in product design and development: A comprehensive review. *International Journal on Interactive Design and Manufacturing (IJIDeM), 18*(6), 3633–3680.
13. Xue, L. (2022). *Designing effective augmented reality platforms to enhance the consumer shopping experiences* (Doctoral dissertation), Loughborough University.
14. Moorhouse, N., tom Dieck, M. C., & Jung, T. (2018). Technological innovations transforming the consumer retail experience: A review of literature. *Augmented Reality and Virtual Reality: Empowering Human, Place and Business*, 133–143.
15. Wang, J., Sun, Y., Zhang, L., Zhang, S., Feng, L., & Morrison, A. M. (2024). Effect of display methods on intentions to use virtual reality in museum tourism. *Journal of Travel Research, 63*(2), 314–334.
16. Baltierra, S. (2023, January). *Virtual reality and augmented reality applied to E-Commerce: A literature review.* Human-Computer Interaction: 8th Iberoamerican Workshop, HCI-COLLAB 2022, Havana, Cuba, October 13–15, 2022, Revised Selected Papers. Springer Nature, p. 201.
17. Plotkina, D., Dinsmore, J., & Racat, M. (2022). Improving service brand personality with augmented reality marketing. *Journal of Services Marketing, 36*(6), 781–799.
18. Peukert, C., Pfeiffer, J., Meißner, M., Pfeiffer, T., & Weinhardt, C. (2019). Shopping in virtual reality stores: the influence of immersion on system adoption. *Journal of Management Information Systems, 36*(3), 755–788.
19. Halid, H., Ravesangar, K., Mahadzir, S. L., & Halim, S. N. A. (2024). Artificial Intelligence (AI) in Human Resource Management (HRM). In *Building the future with human resource management* (pp. 37–70). Cham: Springer International Publishing.
20. Syed, A. A., Gaol, F. L., Pradipto, Y. D., & Matsuo, T. (2021). Augmented and virtual reality in e-commerce—A survey. *ICIC Express Letters, 15*, 1227–1233.
21. Walentek, D., & Ziora, L. (2023). A systematic review on the use of augmented reality in management and business. *Procedia Computer Science, 225*, 861-871.
22. Carton, S. (2019). *What impact will immersive technologies such as augmented and virtual reality have on the retail sector?* (Doctoral dissertation, Dublin, National College of Ireland).
23. Soliman, M., & Al Balushi, M. K. (2023). Unveiling destination evangelism through generative AI tools. *ROBONOMICS: The Journal of the Automated Economy, 4*(54), 1.
24. Mohamed, Ů. (2024). Integrating digital techniques/technologies in developing Egyptian museums (Case study: Alexandria library museums-Alexandria City). *Sohag Engineering Journal*, 4(1), 34–47.
25. Bhatnagar, S. (2022). *Digital disruptions and transformation of bank marketing.* NIBM Working Paper Series, WP 1/April.

26. Singh, B. (2023). Blockchain technology in renovating healthcare: Legal and future perspectives. In *Revolutionizing healthcare through artificial intelligence and internet of things applications* (pp. 177–186). IGI Global.
27. Henke, N., & Bughin, J. (2016). *The age of analytics: Competing in a data-driven world.* McKinsey Global Institute.
28. Kaushal, V., & Yadav, R. (2023). Learning successful implementation of chatbots in businesses from B2B customer experience perspective. *Concurrency and Computation: Practice and Experience, 35*(1), e7450.
29. Singh, B. (2023). Federated learning for envision future trajectory smart transport system for climate preservation and smart green planet: Insights into global governance and SDG-9 (Industry, Innovation and Infrastructure). *National Journal of Environmental Law, 6*(2), 6–17.
30. King, K. (2022). *AI strategy for sales and marketing: Connecting marketing, sales and customer experience.* Kogan Page Publishers.
31. Savickaite, S. (2024). *Using virtual reality to explore individual differences in perception due to neurodiversity* (Doctoral dissertation), University of Glasgow.
32. Sadowski, J. (2023). Total life insurance: Logics of anticipatory control and actuarial governance in insurance technology. *Social Studies of Science.* Advance online publication. https://doi.org/10.1177/03063127231186437
33. Sharma, A., & Singh, B. (2022). Measuring impact of e-commerce on small scale business: A systematic review. *Journal of Corporate Governance and International Business Law, 5*(1), 34–38.
34. Gao, Y., & Liu, H. (2022). Artificial intelligence-enabled personalization in interactive marketing: A customer journey perspective. *Journal of Research in Interactive Marketing, 17*(5), 663–680.
35. Rajamannar, R. (2021). *Quantum marketing: Mastering the new marketing mindset for tomorrow's consumers.* HarperCollins Leadership.
36. Singh, B. (2022). Understanding legal frameworks concerning transgender healthcare in the age of dynamism. *Electronic Journal of Social and Strategic Studies, 3*, 56–65.
37. Ancillai, C., Terho, H., Cardinali, S., & Pascucci, F. (2019). Advancing social media driven sales research: Establishing conceptual foundations for B-to-B social selling. *Industrial Marketing Management, 82*, 293–308.
38. George, A. S. (2023). Future economic implications of artificial intelligence. *Partners Universal International Research Journal, 2*(3), 20–39.
39. Wheeler, J. (2023). *The digital-first customer experience: Seven design strategies from the world's leading brands.* Kogan Page Publishers.
40. Sedkaoui, S. (Ed.). (2018). *Big data analytics for entrepreneurial success.* IGI Global.
41. Neuwirth, R. J. (2023). Prohibited artificial intelligence practices in the proposed EU artificial intelligence act (AIA). *Computer Law & Security Review, 48*, 105798. https://doi.org/10.1016/j.clsr.2023.105798
42. Schaeffer, E. (2017). *Industry X.0: Realizing digital value in industrial sectors.* Kogan Page Publishers.

43. Singh, B. (2019). Profiling public healthcare: A comparative analysis based on the multidimensional healthcare management and legal approach. *Indian Journal of Health and Medical Law, 2*(2), 1–5.
44. Venkatesh, D. N. (2021). *Winning with employees: Leveraging employee experience for a competitive edge.* SAGE Publishing India.
45. Maes, P. (2018). *Disruptive selling: A new strategic approach to sales, marketing and customer service.* Kogan Page Publishers.
46. Berman, B., & Pollack, D. (2021). Strategies for the successful implementation of augmented reality. *Business Horizons, 64*(5), 621–630.
47. Bug, P., & Bernd, M. (2020). The future of fashion films in augmented reality and virtual reality. In *Fashion and film: Moving images and consumer behavior* (pp. 281–301). Springer Singapore.
48. Deshbhratar, S., Joshi, S., Alwaali, R. N., Saear, A. R., & Marhoon, H. A. (2023, September). Augmented reality of online and physical retailing: A study of applications and its value. In *AIP conference proceedings* (Vol. 2736, No. 1). AIP Publishing.
49. Meißner, M., Pfeiffer, J., Peukert, C., Dietrich, H., & Pfeiffer, T. (2020). How virtual reality affects consumer choice. *Journal of Business Research, 117,* 219–231.
50. Semerádová, T., & Weinlich, P. (2022). The place of virtual reality in e-retail: Viable shopping environment or just a game. In *Moving businesses online and embracing e-commerce: Impact and opportunities caused by COVID-19* (pp. 92–117). IGI Global.

Index

For Product Safety Concerns and Information please contact our EU representative GPSR@taylorandfrancis.com Taylor & Francis Verlag GmbH, Kaufingerstraße 24, 80331 München, Germany